Teaching Mathematics in Early Childhood

Teaching Mathematics in Early Childhood

by

Sally Moomaw, Ed.D.
University of Cincinnati
Ohio

Baltimore • London • Sydney

Paul H. Brookes Publishing Co.
Post Office Box 10624
Baltimore, Maryland 21285-0624
USA

www.brookespublishing.com

Book design by Erin Geoghegan.
Typeset by Aptara, Inc., Falls Church, Virginia.
Manufactured in the United States of America by
Sheridan Books, Inc., Chelsea, Michigan.

The cover photograph and the photographs in this book are used by permission of the individuals pictured or their parents and/or guardians.

Cover and interior photographs (except photograph on p. 161) by Heather L. Guthridge.

Photograph on p. 161 by Sally Moomaw.

In most examples, names have been changed to protect identity; actual names are used in a few examples by permission of the individuals themselves or their parents and/or guardians.

Library of Congress Cataloging-in-Publication Data

Moomaw, Sally
  Teaching mathematics in early childhood / Sally Moomaw.
      p.   cm.
    Includes bibliographic references and index.
    ISBN-13: 978-1-59857-119-6
    ISBN-10: 1-59857-119-2
    1. Mathematics—Study and teaching (Early childhood) I. Title.

QA135.5.M6152 2011
372.7—dc22   2010053199

British Library Cataloguing in Publication data are available from the British Library.

2015   2014   2013   2012   2011

10   9   8   7   6   5   4   3   2   1

Contents

About the Author

Sally Moomaw, Ed.D., is Assistant Professor of Early Childhood Education at the University of Cincinnati in Ohio. She taught preschool and kindergarten children in inclusive, diverse classrooms for more than 20 years. She is the author or coauthor of 13 books for early childhood education, including *More than Counting: Whole Math Activities for Preschool and Kindergarten* (Redleaf Press, 1995), *More than Magnets: Exploring the Wonders of Science in Preschool and Kindergarten* (Redleaf Press, 1997), *Lessons from Turtle Island: Native Curriculum in Early Childhood Classrooms* (Redleaf Press, 2002), and *More Than Counting: Standards Edition* (Redleaf Press, 2011). She has given numerous presentations for educators throughout the United States and has developed a Mathematics Toolkit for the Ohio Department of Education to help preschool and kindergarten teachers implement state content standards. Her research focus is the development of mathematics understanding in young children.

Preface

The focus of this book is standards-based mathematics education in preschool and kindergarten. Designed to link theory and research to effective pedagogical practice, the book specifically targets the development of mathematical reasoning in young children and guides teachers in creating and implementing curriculum and instructional procedures to maximize conceptual understanding. The publication of the National Council of Teachers of Mathematics' *Principles and Standards for School Mathematics* (2000) and the subsequent adoption of preschool and kindergarten mathematics standards by many states have focused the attention of early childhood educators on the depth and breadth of mathematical learning that should be addressed in preprimary education. The preschool and kindergarten years are now viewed as critical to later school success in mathematics. This book is designed to move prospective and practicing teachers from a developmental understanding of each mathematics content standard through the creation, implementation, and assessment of curriculum.

Preschool and kindergarten teachers increasingly have the opportunity to work with children of varying abilities and challenges within inclusive environments. For this reason, this book also targets early intervention. After describing the typical developmental trajectory of key mathematical concepts, it addresses curriculum accommodations and instructional strategies to support children with disabilities. Universal design for learning is an important topic addressed throughout the book.

Teachers of young children work within a wide variety of settings. Some have a prescribed mathematics curriculum, and others create their own curricular activities within an overall framework. This book is designed to support teachers in both types of settings through a curriculum that encompasses individual and small-group activities, large-group activities, and integration throughout the curriculum. Teachers can create a standards-based curriculum modeled after the numerous sample activities presented in the book, or select various types of activities to augment an adopted curriculum. Throughout the book, an emphasis is placed on the importance of engaging children in conversations about mathematics, or "math talk," as well as the importance of embedding mathematical situations within play and daily living scenarios.

Most of the activities in this book were designed for preschool and kindergarten children, age 3 years and older. Some of the activities call for materials containing small pieces. Teachers who have children who still put objects into their mouths should be careful to use materials that cannot be swallowed.

REFERENCE

National Council of Teachers of Mathematics. (2000). *Principles and standards for school mathematics.* Reston, VA: Author.

Acknowledgments

I would like to thank the following people: Charles Moomaw, for his computer technical support throughout the preparation of this manuscript; Peter Moomaw, for sharing his mathematical expertise; Heather L. Guthridge, for her photographic work; and editor Astrid Zuckerman, for her enthusiasm and support for this project. Special thanks go to Tehya, Chanda, and Evan for allowing us to photograph them as they participated in the activities.

To Charlie

Foundations for Mathematical Development

Mathematics is a human activity, a social phenomenon, a set of methods used to help illuminate the world, and it is part of our culture.

—Jo Boaler (2008, p. 16)

> *"My puppy wants to say good-bye to the horses," announced 3-year old Elena, whose family was getting ready to leave the farm. She held her stuffed animal up to the nose of a large horse, which was looking out of its stall window.*
>
> *"Okay," said Elena's mom. "Now climb into your car seat."*
>
> *"No!" protested Elena. "What about those other two horses?" She remembered that two other horses were stalled on the other side of the barn, out of view. Although still quite young, Elena had counted three horses in the barn and mentally separated that number into the one horse she could see and the two horses she could not see. In essence, she had subtracted one from three.*

Young children develop mathematics concepts long before they enter school. They use this informal math knowledge to mentally organize their environment, make sense of daily life experiences, and expand their play. Young children like mathematics. They practice saying counting words, compare small numbers of items, and use their fingers to represent small numbers. Young children play with mathematics. They stack blocks in towers of ascending size, make sure that each teddy bear gets a sip of pretend juice, and diligently experiment with fitting shapes into matching holes. Young children challenge themselves with mathematics. They attempt to add and compare scores in games, try to figure out how many more toy cars they need in order to have the same number of cars as their friend, and attempt to divide cookies fairly among a group. Yet, despite this capability and interest in mathematics, by early elementary school many children say that they hate math. Their enjoyment of mathematics and confidence in their abilities are gone. What has happened to these children? Obviously, the years between age 3 and the beginning of more formal education are critically important.

For young children to develop optimally, preschool and kindergarten curricula need to have a strong emphasis on mathematics. Both the National Council of Teachers of Mathematics (NCTM) and the National Association for the Education of Young Children (NAEYC) view early childhood as vital for children's mathematical development (Baroody, Lai, & Mix, 2006). Recent research indicates that school-entry math skills have the greatest predictive power for later achievement in school, compared with reading, attention skills, and social development (Duncan et al., 2007). Unfortunately, research also shows that preschool children do not have much exposure to mathematically oriented activity in either home or formal child care settings (Tudge & Doucet, 2004). Even in kindergarten, children receive, on average, only half as much instruction in math as in reading (Hausken & Rathbun, 2004).

Expectations regarding the curriculum that children should experience during preschool and kindergarten have changed markedly since the adoption of content standards at the national and state levels. NCTM was the first national organization to respond to the call for national standards in education. Its *Curriculum and Evaluation Standards for School Mathematics* (NCTM, 1989) set specific goals for educators to improve mathematics education and focused heavily on thinking and problem-solving skills. This document was followed by *Principles and Standards for School Mathematics* (NCTM, 2000), which integrated curriculum, teaching, and evaluation. In producing these seminal documents, NCTM included preschool education as the foundation for children's mathematical development. The standards outlined by NCTM have framed the mathematics standards adopted by many states across the country.

NAEYC also acknowledges the importance of early mathematics education (Tomlinson & Hyson, 2009). Emphasis is placed on the need for children to develop mathematical relationships, problem-solving skills, and reasoning. A pioneering effort in this regard was the NAEYC publication of *Number in Preschool and Kindergarten* (Kamii, 1982), which reframed teaching strategies in mathematics for preschool and kindergarten teachers by emphasizing daily living situations and games in order to stimulate children's numerical thinking. Since that work was released, much research has been published regarding the teaching and learning of mathematics in early childhood education (Clements & Sarama, 2007). Important research studies, along with the documented reflective practice of highly effective early childhood mathematics teachers (Andrews & Trafton, 2002; Baratta-Lorton, 1976; Moomaw & Hieronymus, 1995, 1999; Schielack & Chancellor, 2010), provide guidance for preschool and kindergarten teachers to become successful mathematics educators.

MATHEMATICS IN THE PRESCHOOL AND KINDERGARTEN CLASSROOM

Mathematics in preschool and kindergarten classrooms begins with play. As young children interact with one another, mathematical situations develop naturally. Perhaps Max is busy setting the table and wants to make sure that each person has a plate, a cup, and a spoon. This action demonstrates his comprehension of a **one-to-one correspondence** relationship, the understanding that each object in one group can be paired with one object from a second group, a foundational mathematics concept. Then Maria brings a bowl of plastic eggs to the table. Noticing the plates that Max has set out, she deals out the eggs, one at a time, until each plate has three eggs. This action illustrates informal knowledge of division, the idea that a group can be divided equally into a specific number of parts. The teacher joins the play. "I'm going to feed this baby an egg for breakfast," she says.

"How many eggs do I have left?" She now has introduced a new idea, subtraction, into the play.

This familiar scenario illustrates an important aspect of successful mathematics teaching in the early years; that is, although informal mathematics occurs regularly in children's play, learning advances when teachers are prepared to interject mathematical situations into areas of the children's interest. This extension of informal knowledge to include more formal mathematical language is critical. When the informal mathematical knowledge of preschool and kindergarten children is not connected to the formal learning they are likely to receive in the primary grades, learning gaps can occur that are difficult to bridge (Hatano, 2003). Also, children enter preschool and kindergarten with different amounts of informal mathematical knowledge. This gap in understanding needs to be addressed before the primary grades so that children will have a conceptual foundation to support later learning (Baroody et al., 2006). Skillfully injecting mathematics into play situations is critical because young children typically do not respond to instruction or answer questions that do not interest them. Questioning or modeling that evolves naturally from children's initiatives captures their attention and causes them to think about new concepts and relationships that are within their reach.

Integrating Mathematics Learning

Although children can and do learn from play experiences, teachers cannot assume that mathematical learning will automatically unfold. It must be planned. First, teachers themselves must understand the important mathematical concepts that children should construct during the preschool and kindergarten years. With these always in mind, teachers can design the classroom environment to support this learning. Materials that are highly likely to inspire mathematical situations can be carefully assembled and placed in centers throughout the classroom. Then, as teachers monitor the ensuing play, they can interject comments or model scenarios that introduce, support, or extend mathematical concepts. The examples that follow illustrate how this process might unfold.

EXAMPLE 1.1

Luis's full-day summer kindergarten class includes children with a wide developmental range. Most of the children have strong counting skills and are beginning to add small numbers; however, some of the younger children still skip over objects when counting, and one child with cognitive delays uses one-to-one correspondence to quantify. Luis decides that a post office in the dramatic play area of the classroom would be interesting to the children and would generate many opportunities for mathematical learning.

To accommodate children who are beginning to add, Luis uses his computer to design stamps with values ranging from 1 to 5 cents and prints them on labels. He plans to include pennies and a cash register in the center so that children can sell stamps to one another. He expects these experiences to give children many opportunities to count pennies and add the quantities on the stamps. Luis also includes many inexpensive and recycled envelopes and small packages in the area. In a meeting with his assistant, he suggests that they encourage the child with cognitive delays to find one stamp for each of a small quantity of envelopes or packages. This will support the child's understanding of one-to-one correspondence and provide opportunities for the teachers or peers to model counting.

In this case, opportunities for learning numerical concepts are not left to chance. The teacher, who believes in the power of a play-based curriculum, carefully designs the environment to support the mathematical concepts that the students are learning.

> **EXAMPLE 1.2**
>
> *Rose's preschool class is fascinated by the construction of an apartment building across the street from the school. The children decide that they would like to build their own town in the block area and describe it in words and pictures in a class book. During the planning stage, Rose helps the children form into small groups and decide on the buildings that they want to build. As each building is constructed and completed, Rose takes digital photographs.*
>
> *After all of the buildings have been completed, Rose meets with the children in groups to help them write their section of the class book. Because Rose wants the children to consider geometry, measurement, and number concepts in their descriptions, she asks each group the following questions that focus on these areas:*
>
> - *What shapes did your group use to build this structure?*
> - *What did you do first to build your building?*
> - *How many cylinders are holding up the floor?*
> - *Where did you put the largest blocks?*
> - *What shapes did you use for decoration on the roof?*

This building process allows Rose's class to focus on geometric forms. The block area is an excellent area of the classroom to support such learning. Because all of the children participated in the project, everyone had the opportunity to explore these concepts.

Mathematics in Small Groups

Another effective way to implement mathematics curriculum in preschool and kindergarten classrooms is through carefully designed small-group activities. Small groups allow the teacher to support each individual child on the basis of his or her level of understanding. Small groups also encourage peers to exchange ideas and model for one another. Often, the activities planned for small groups are math games in which children roll a die and then take a corresponding amount of counters or move a corresponding number of spaces along a path. Experiences in the geometry, algebra, and measurement areas also can be implemented at the small-group level. In preschool, children may self-select into these activities, whereas in kindergarten, small-group experiences may be part of the daily math curriculum.

Mathematics in Large Groups

Large-group experiences provide a venue for teachers to reinforce mathematics concepts through counting songs, books that involve mathematical situations, and rhythm or movement patterns. Teachers also can introduce new concepts or activities, such as a collection to sort by various attributes or a new patterning activity. In addition, group times provide a forum for children to exchange ideas and, perhaps, to vote on topics of interest or importance. Teachers can represent children's votes on bar graphs and lead them in analyzing the results. When individual and small- and large-group experiences are all included as important parts of the preschool and kindergarten curriculum, children encounter mathematics often throughout the day. For this reason, each of these areas is highlighted in the chapters that follow.

STANDARDS IN EARLY MATHEMATICS EDUCATION

The NCTM standards are organized across grade bands, starting with prekindergarten through Grade 2, and are divided into content and process standards (NCTM, 2000). Content standards encompass the knowledge that children are expected to learn across five domains of mathematics, whereas process standards are the means children use to acquire and apply these concepts. Content and process never can be separated, because it is through the process of learning that children come to know and understand content. Nevertheless, for purposes of discussion, the two types of standards are grouped according to content or process. They are equally important.

Mathematics Content Standards

There are five mathematics content standards: Number and Operations, Algebra, Geometry, Measurement, and Data Analysis and Probability. Each of these standards is explored in depth in the chapters that follow with respect to its impact on teaching mathematics in preschool and kindergarten. Abbreviated descriptions of the content standards follow.

1. *Number and Operations:* This standard focuses on the development of a deep understanding of number and number sense, or the ability to decompose numbers and use them as referents. Even very young children learn that "2" can be decomposed as "1 for me and 1 for you." For young children, Number and Operations includes quantifying small amounts; comparing sets of objects as more, less, or equal; counting; ordering numbers (first, second, last, and so forth); combining sets (early addition); taking away from sets (early subtraction); and dividing materials among friends (early division). It involves understanding underlying relationships, such as one-to-one correspondence (one number word for each object counted) and **cardinality** (the last number counted equals the total). Much of the mathematical learning that occurs through children's play involves Number and Operations.

2. *Algebra:* The Algebra standard involves understanding patterns and relationships. It also includes analyzing, representing, and modeling mathematical situations. In the preschool setting, children construct mathematical relationships by sorting and classifying materials and eventually forming them into patterns. Mathematical situations arise during daily living and play experiences. For example, children might have to decide what to do when three children want to play at the water table and only two spaces are available.

3. *Geometry:* The Geometry standard includes understanding spatial relationships, positional terms, and the properties of two- and three-dimensional objects. In preschool and kindergarten, children learn fundamental geometric concepts through playing with blocks and manipulative materials. For example, they discover that rectangular blocks are good for building floors because they have flat sides and corners. Yet, almost every child experiences the challenge of trying to balance a tall, thin block vertically, only to have it come crashing down. These concrete experiences form the foundation for later analytic processes in geometry. They also encourage reasoning and problem solving and eventually help children form inferences, such as curved solids roll.

4. *Measurement:* Young children are constantly making measurement comparisons. Who has more juice in her glass? How much did I grow since my last birthday? Can I stack the blocks higher than I am? Why do I have to stand on a stool to reach the sink and my brother doesn't? The measurement standard encompasses and extends these early experiences. It involves understanding the measurable attributes of objects, constructing the concept of an appropriate unit of measure, the application of number to measurement,

and measurement comparisons. To **seriate,** or order objects by size, is an aspect of measurement frequently explored by young children and included under this standard.

5. *Data Analysis and Probability:* Young children are interested in concepts of data analysis and probability as they apply to their daily lives. How many children rode the bus to school? Are there more children who like vanilla or chocolate ice cream? Are there enough mittens in the bin for every child to have a pair? Which fruits received the most votes for fruit salad? What is the probability that it will rain and curtail outside play? Dealing with important data encourages young children to think and problem solve while applying concepts related to number and measurement. For example, they might help graph the fruit votes and decide which ones have the most votes by looking for the highest columns or by counting the votes.

Mathematics Process Standards

There are also five mathematics process standards: Problem Solving, Reasoning and Proof, Communication, Connections, and Representation. These process standards encompass the means that children use to learn mathematical concepts, as well as their methods for communicating results and describing their reasoning. Mathematical processes are critical because they reveal the thinking that underlies mathematical understanding. A focus on content that excludes an equal focus on process leads to students' having a surface knowledge of mathematics. Although they may have memorized a substantial amount of information, students may be unable to apply it to unique mathematical situations. This shortcoming becomes evident as students advance in mathematics. They soon reach a point at which a solid understanding of content, as provided by the development of critical processing skills, is essential. Fear of higher level mathematics, or math phobia, is directly related to a lack of processing and application skills. Students' learning in mathematics is thus inhibited by an insufficient focus on processing standards in their education. The five process standards are described as follows:

1. *Problem Solving:* Problem solving involves drawing on previous knowledge to solve a unique problem. It is a primary means by which children develop mathematical understanding. Important learning occurs when adults allow children to solve problems. This does not mean that adults offer no help; however, rather than telling children how to solve a problem, skillful teachers help them frame the problem and then ask questions or supply comments to help them access and apply what they already know. For example, a child might want to know how many "google eyes" she needs for her picture of three people. Rather than supply the answer, the teacher might say, "How many eyes does each person have?" By isolating this part of the problem, she will likely help the child break the problem down into meaningful steps and arrive at a solution. As in the previous example, for young children, problem solving emerges through play and daily experiences. Skillful educators intentionally take advantage of these situations in order to develop independent, flexible thinkers.

2. *Reasoning and Proof:* Reasoning is central to mathematical understanding. Under this standard, children are expected to develop and evaluate mathematical arguments. As with other areas of mathematics, for young children this often emerges through play and daily life experiences. For example, a child might accuse another of taking too many orange slices. This situation poses a nice opportunity to emphasize reasoning and proof. All the adult needs to ask is, "How do you know?" or "What makes you think so?" Both children can then discuss their own reasoning and demonstrate a proof. They may count the orange slices or perhaps align them in rows to make a comparison. Either way, they are solidifying mathematical concepts and strengthening their logical thinking.

3. *Communication:* The Communication standard emphasizes the importance of asking children to formulate and express their mathematical thinking. In doing so, they may discover that they made a mistake and self-correct, or a peer may point out a discrepancy. Communication encourages children to use and understand the language of mathematics, which is crucial for later mathematical learning. By testing their ideas on others, children validate, refine, or change their thinking. Open communication can occur only in an environment that emphasizes the importance of the thinking process as opposed to the need for correct answers. Young children regard mathematics as fun and exciting when they can approach it as a puzzle that is ripe for solving. Skillful educators help children understand that there are many different ways to solve a mathematical problem. This enables children to apply their own understanding to problem situations, rather than rely on an adult for the one correct way to proceed.

4. *Connections:* Although the mathematics content standards are separated into different areas, they are all related and interconnected. As children begin to recognize and understand these connections, their understanding of mathematical concepts deepens. A key role for the teacher is to help children make these connections. For example, at snack time, the teacher might ask the children how many different ways they can find to create "3" with their mixture of colored Goldfish crackers. This would help children connect concepts of number to algebraic reasoning. Or the teacher might ask whether a row of three grapes is as long as a row of three carrot sticks. This would help children connect number with measurement concepts. Helping children discover that a triangle has three sides, whereas a square and a rectangle each have four sides, is yet another example.

5. *Representation:* The Representation standard acknowledges that there are many different ways to represent mathematical concepts and that children should be encouraged to use a variety of methods to represent and communicate mathematical ideas. Children might represent a mathematical problem with physical objects, such as constructing a train with a specific number of cars and passengers. Drawing would be another method for children to model the same problem. Some children might use their fingers, with one hand indicating cars and the other hand indicating passengers. Still other children might create hash marks to quantify the two groups. Some children might use a numeric symbol for each group. Observing the wide variety of ways that their peers represent mathematical problems helps children to solidify their own understanding and develop mathematical connections.

DEVELOPMENTALLY APPROPRIATE MATHEMATICS CURRICULUM

When and how curriculum is implemented is as important as the content. Young children cannot grasp concepts that are too far above their level of understanding, and they quickly lose interest in activities that are not engaging. NAEYC considers developmentally appropriate practice in mathematics teaching to include 1) opportunities for talking about mathematics during daily classroom experiences, 2) the implantation of mathematics experiences through individual and small- and large-group experiences, 3) an emphasis on reasoning and problem solving, 4) the application of research-based knowledge regarding developmental progressions in children's understanding of math concepts, and 5) a focus on the major content areas of mathematics (Copple & Bredekamp, 2009).

The Importance of Math Talk

Research shows that the amount of math-related talk that teachers provide is significantly related to children's mathematical learning (Klibanoff, Huttonlocher, Vasilyeva, & Hedges,

2006). Through their conversations with children, teachers can introduce mathematics vocabulary within interesting and meaningful contexts. In addition, through play experiences, they can interject the types of mathematical questions that children will need to understand in later, more formal educational settings.

Many interactions with preschool and kindergarten children contain the potential for math talk. It is unfortunate that most of these opportunities are missed. Many teachers are not comfortable talking about math, do not enjoy math, or do not recognize the potential for math-related discussions that exist in their classrooms. With preparation and practice, however, this tendency to avoid math-related conversations can change.

When adults attend parties, they often try to think of conversation openers. The weather, work, vacation plans, or sports might be selected to fill this need. Likewise, if teachers want to inject math talk into their conversations with children, it may be helpful to plan conversation openers. A few examples follow:

- Molly, I see clips in your hair today that have butterflies and bows. Do you think you have the same number of butterflies as bows? Let's look in the mirror.

- Reggie brought a toy truck to school today. I wonder how many wheels it has. Do we get to count the steering wheel, too?

- It's really cold today. How many of you forgot to wear mittens? I'll get the extra mitten bin. If four children need mittens, how many mittens do I need to get out of the bin?

- I'm going to order a new set of books for our class from the library. If I ask for two books for each child, how many will that be? I need to tell the librarian.

- I see some farms being built in the block area. Do we have enough animals for each of you to have four? How can we find out?

- How many grapes do you want for snack? Four? Here's a spoonful. Check to make sure I gave you enough grapes.

The list could go on and on. Because all of the questions relate directly to children's interests or needs, they are likely to result in responses from the children and may lead to a more extended conversation. Math talk is a recurring focus of this book.

Individual and Group Math Experiences

Most preschool and kindergarten classrooms include children of varying ages, developmental levels, and experiences. This is particularly true in multi-age groupings, which are common in preschool programs. For this reason, activities for young children are often planned with individuals in mind. For example, a classroom might include several children who are beginning to count and represent small quantities of objects. For these children, the teacher would want to include games with a die or spinner with 1–3 dots to support foundational counting concepts. Children could use these materials individually or in small groups. In the same classroom, other children might be transitioning into addition and be ready to add two dice with 1–6 dots when playing games. A well-planned curriculum can accommodate children easily at different conceptual levels of understanding.

Large-group experiences also can meet the needs of a range of students. In a classroom such as the one just described, the teacher might plan an extended counting activity—for instance, Ten in the Bed, in which one person is removed from a "bed" for each verse of the song. In order to determine how many are left in the bed, a re-counting of the remaining people is necessary after each verse. An activity such as this gives children who are strong at forward counting the opportunity to count backward and introduces the concept of subtraction by one. It also supports children who are learning to count forward because the remaining people repeatedly are counted in this way. Children who can count up to three see counting of higher quantities modeled repeatedly in an interesting contextual situation.

Development of Mathematical Thinking

All of the process standards should be integrated into mathematics learning. To help teachers understand how the process standards relate to learning in each of the mathematics content standards, the process standards are highlighted in a special section of Chapters 2–7 of this book. Teachers of young children often use comments and questions to encourage children to communicate, draw connections between mathematics domains and other areas of the curriculum, and represent and explain their responses to problem-solving situations. For that reason, many examples of this type of conversation are included throughout the book.

Development of Mathematical Concepts

Children often develop mathematical concepts in a predictable sequence, although the ages at which the children demonstrate particular concepts and the speed at which they learn them vary. The areas of number sense, geometry, and measurement have received a great deal of attention by researchers who study children's conceptual development. It is important for teachers to understand and apply this knowledge to their teaching so that they can support and maximize children's learning opportunities. Although the areas of algebra and data analysis have received less attention from researchers, skilled early childhood teachers have explored curricula in these areas in their classrooms and published their reflections. Knowledge about children's development in each of the mathematics content areas is discussed in the chapters that follow and is then tied directly to curriculum planning and implementation.

Mathematics Focus Areas

Three domains of mathematics—number and operations, geometry, and measurement—have been designated as focus areas in preschool and kindergarten by NCTM, NAEYC, and the National Research Council (NRC) (Copple & Bredekamp, 2009; NCTM, 2006; NRC, 2009). These areas are targeted because they are mathematically important, have a research base that supports young children's learning trajectories, and connect with learning at later grade levels. Algebra and data analysis, however, are recognized as supportive of learning in the focus areas, as well as being directly connected to them. In contrast, the National Mathematics Advisory Panel (2008) recommends that all early mathematics instruction lead toward the development of algebra concepts, although it acknowledges that fluency with numbers and certain aspects of geometry and measurement are important foundations for later algebra.

Instruction in all five content areas has merit in preschool and kindergarten. Although number sense is central to the mathematics curriculum in the early grades, the other content areas are also important. All five content areas are closely connected. For example, Moomaw and Hieronymus (1995) describe group graphing, which is part of the Data Analysis and Probability standard, as an important part of the curriculum because it provides opportunities for children to create and compare sets on topics that are of interest to them, such as voting responses. Patterning, a primary component of the Algebra standard, helps children develop, understand, and connect important mathematical relationships, such as symmetry in geometry and seriation in measurement. For these reasons, preschool and kindergarten teachers should plan connected mathematics curricula in all five content areas.

MATHEMATICS IN INCLUSIVE CLASSROOMS

Although most preschool and kindergarten classrooms include children at various developmental levels and with varying instructional needs, many schools now include children with diagnosed special needs in general education classrooms. This practice is termed **inclusive education.**

"Inclusion is a philosophy that brings diverse students, families, educators, and community members together to create schools and other social institutions based on acceptance, belonging, and community" (Salend, 2008, p. 5). Inclusive classrooms require teachers to adopt a wider range of instructional practices than they may have used in the past in order to accommodate the needs of all students. This focus on extended and enhanced teaching practices is frequently beneficial to all students, particularly in mathematics, an area in which many typically developing children often struggle. The platform for designing and implementing a curriculum that accommodates all students is referred to as **universal design for learning (UDL).**

Universal Design for Learning

The concept of universal design originated in the field of architecture, in which it was determined that designing structures in the first place to accommodate all users was more efficient and less costly than redesigning them later to comply with equal-access laws. The underlying concepts of universal design have been widely adopted in education to refer to curriculum, teaching strategies, and environments that support all learners, including those with specific disabilities (Salend, 2008).

There are three core principles in UDL:

1. *Multiple means of representation:* Instruction should provide many ways for students to acquire knowledge.

2. *Multiple means of expression:* Students should have many different ways to demonstrate what they know.

3. *Multiple means of engagement:* Educators should employ many different methods to motivate learners.

These principles of instruction are readily applicable to mathematics instruction for all children in preschool and kindergarten. First, all young children need curriculum and instruction that build on their informal knowledge of mathematics. This requires that math concepts be imbedded in play and daily living experiences. Some of the many instructional strategies that help all children learn are modeling math concepts with objects that are directly linked to children's literature; singing, counting, adding, and subtracting songs; performing movements and rhythms that accentuate patterns; sorting objects according to many attributes, including texture; playing mathematics games; and extending math activities into multiple areas of the curriculum, including the sensory table, blocks, art, and music.

Young children also have many different ways of demonstrating what they know about mathematics. For example, they may represent one-to-one correspondence in their play by giving one bottle to each doll baby. Children may line up toy frogs next to toy turtles in the water table to determine which group has more. They may use hash marks or beads on an abacus to keep track of their score in a gross-motor game or create gestures to describe large versus small objects. Children may show their conceptual understanding of geometric forms, balance, and symmetry through structures they create in the block area. These are some of the many ways that children demonstrate mathematical understanding in preschool and kindergarten classrooms and that teachers can use to extend learning.

Teachers of young children quickly learn that activities must be interesting or children will not attend. For this reason, effective teachers quickly learn to present concepts in ways that engage young children. Games, songs, books, fingerplays, movement, art, and dramatics are some of the many ways that preschool and kindergarten teachers can engage all children in mathematics learning.

Although the range of activities introduced by skillful early childhood teachers accommodates the learning needs of most learners, some children require modifications to particular activities or experiences in order to actively participate. These modifications should be made *before* the curriculum is implemented so that the full community of learners can participate. Various areas of mathematics may present particular challenges to children with certain types of disabilities. For this reason, every curriculum chapter in this book highlights UDL with specific children in mind.

Multileveling activities, in which various children can participate in different ways according to their level of development, are particularly effective in building a community of learners and peer instructors. This occurs when children who are more advanced in their thinking model and explain their strategies to younger, less experienced, or less advanced peers. These experiences are valuable for all children because younger children are not intimidated by opinions offered by older children, and more advanced students must think through the reasoning behind their own mathematical thinking in order to explain it to other children. Most of the curriculum activities presented in the rest of the chapters meet the needs of children at multiple levels of understanding and can be readily extended to be easier or more difficult.

WHAT IT MEANS TO BE A MATHEMATICS TEACHER

Many preschool and kindergarten teachers do not think of themselves as mathematics teachers, although most are comfortable teaching literacy. A recent, informal survey of college seniors in an early childhood classroom revealed that they all felt competent and excited about teaching reading but that only 10% felt the same way about mathematics. This may be partially the result of state mandates for schools of education that focus heavily on early literacy development (NRC, 2009). It probably also reflects prevailing attitudes among early childhood teachers about mathematics content and instruction. Many of these teachers are uncomfortable with mathematics, with math anxiety often dating to their own elementary school experiences, in which an emphasis was placed on correct answers and insufficient time was devoted to word problems and problem solving (Philipp, 2007).

Preschool and kindergarten teachers are expected to be generalists; in other words, they must have sufficient knowledge and instructional skills to teach all of the content areas. This includes mathematics. Although early childhood teachers certainly do not have to be mathematicians, they need to know the relevant content and standards that they are expected to teach. Equally important, teachers need to develop and convey an enjoyment of mathematics. Teachers may find that the same curriculum materials and teaching strategies that interest young children also ignite their own enjoyment of mathematics.

To effectively teach mathematics to young children, teachers need comprehensive knowledge in four areas: 1) characteristics of young children, 2) cognitive learning theories, 3) key principles of mathematics teaching, and 4) mathematics content.

Characteristics of Young Children

Young children are active learners. They learn best when they can manipulate objects as part of their play, interact with other children, and use multiple senses to support their investigations. By manipulating materials, children construct many different kinds of mathematical relationships, such as same/different/similar, more/less/equal, large/medium/small, and so forth. They may pair objects in a one-to-one correspondence relationship, such as one driver for each car, or create simple patterns, such as two animal babies for each parent. These concepts that children construct through their play are foundational for later mathematical learning. Interactions with other children are another critical part of learning. One child may make comments that conflict with something another child believes, such as declaring that two objects are still the same length even though they have been moved into different positions. When children play games together, one may point out a counting error that another has made or model a different strategy for solving a problem. This causes the children to reexamine their own thinking and moves them forward in their development. Finally, children use sensory input to develop cognitive understanding. Some children may respond best to learning in a particular domain. For example, many children extend and repeat patterns first in the music area.

Young children are also autonomous (Erikson, 1950). If they are not interested in a particular activity or experience, they may refuse to participate. In general, young children have shorter attention spans than older students, although some can spend a long time focusing on an activity that they find intriguing. Developing a mathematics curriculum that is interesting to young children and holds their attention is a critical component of mathematics instruction.

The preschool and kindergarten years are a period of rapid development for children. Many progress quickly in their understanding of mathematical concepts, such as strategies for quantifying and comparing sets. For that reason, teachers need to assess children's understanding continually through the children's interactions with materials and other people. In this way, teachers can offer comments or ask questions that maximize children's growth in mathematics.

Cognitive Learning Theories

Two important theorists, Jean Piaget and Lev Vygotsky, have contributed significantly to an understanding of how young children learn. Both are considered to have been constructivists: They believed that individuals construct knowledge from their experiences. Although Piaget's theory has had a broad effect on education in general, some scholars believe that the application of his theory is particularly important in the area of mathematics education:

> All too frequently even many developmentally advanced ("bright") students are *more or less* permanently handicapped by such [nonconstructivist] teaching methods . . . that focus on direct transmission from teacher to student and on correct answers rather than on autonomous thinking and construction of mathematical principles. (Wadsworth, 1989, p. 187)

Aspects of Piagetian Theory   Piaget believed that play is an important vehicle for learning and that children construct their own knowledge by interacting with people and objects in their environment (Mooney, 2000). Piaget posited three types of knowledge: physical, logical–mathematical, and social (Kamii, 1982; Wadsworth, 1989). Understanding what these three types of knowledge mean and how they are constructed by the child has important implications for mathematics instructors.

- *Physical knowledge:* This type of knowledge is rooted in the physical properties of objects: color, texture, weight, shape, and so forth. Children develop physical knowledge by using their senses to actively explore objects. They look, feel, listen, drop, throw, taste, roll, push, and so forth to discover how objects will react. Piaget maintained that a full understanding of physical knowledge could not come simply from listening to someone talk about things or read about them; rather, the child must actively explore the objects.

- *Logical–mathematical knowledge:* This type of knowledge is constructed internally as children reflect upon their interactions with objects and form important relationships. Logical–mathematical knowledge is, in essence, invented by the child on the basis of these actions. Examples of logical–mathematical relationships include more/less/same, three, and one more. Consider the concept of three, which a child might associate with three balls. There is nothing intrinsic in any of the individual balls to denote three-ness. It is only when each ball is considered in relationship to the other two balls that the concept of three can be applied to the balls. Although the concept of number is an internal construction, it is based on the child's exploration of real objects. This is why concrete objects are routinely used in early mathematics experiences. By putting objects into many different relationships, children construct fundamental mathematical concepts.

- *Social knowledge:* The third type of knowledge is socially transmitted knowledge, which involves information based on social norms or customs and that is, therefore, arbitrary. Labels for objects, values, rules, and cultural customs are examples of social knowledge.

According to Piaget's theory, most of mathematics involves logical–mathematical relationships, which cannot be socially transmitted; however, historically, mathematics often has been taught as though the knowledge could be told to the students, who would then understand it (Wadsworth, 1989). Despite concerted efforts to revise teaching models, many teachers continue to follow a transmission-of-information model. Constructivists argue that this is inappropriate because students must construct logical–mathematical relationships through their own thinking. The role of teachers is to provide opportunities for students to think about mathematics in many different ways, thereby facilitating their internal construction of mathematical concepts.

Aspects of Vygtotskian Theory   Vygotsky's theoretical work often is referred to as social constructivism. Although Vygotsky shared Piaget's view of the importance of play to children's cognitive development, he focused heavily on the role of speech in mediating learning, as well as the central influence of culture (Cobb, 2007; Vygotsky, 1978). Vygotsky is perhaps most recognized for his description of the role that social interactions with knowledgeable others plays in intellectual development. From his research with children, Vygotsky found that children of the same chronological age varied greatly in their abilities to learn under equivalent teacher guidance. It was children who were mentally close to understanding the concepts who benefited from the instruction. Vygotsky termed the distance between actual mental development and potential development as the **zone of proximal development (ZPD).** Children who were close to understanding a concept, or who were within this ZPD, could benefit from instruction by an adult or a more competent peer (Vygotsky, 1978).

The widely accepted concept of a ZPD means that teachers must carefully observe children to determine their current thinking level. Only then can they make comments, ask questions, model, or provide other instruction that is just beyond the children's current thinking.

Key Principles of Mathematics Teaching

An overriding principle of mathematics education is that students must understand the mathematics they are learning. The goal is to develop autonomous learners who can apply their understanding of mathematics to the new types of problems they will face in the future (NCTM, 2000). Rather than facing new mathematical situations with trepidation, autonomous students view new problems with confidence, knowing that they can apply the knowledge they already have to new situations, explore a variety of paths to solve new problems, and create new understanding. In order to learn, students must take some initiative. A mathematics professor in a college classroom recently advised his students, "Try something out. Doing something is always better than doing nothing. If you don't try anything out, we won't have anything to talk about" (S. Pelikan, personal communication, May 12, 2010).

In keeping with the goal of developing autonomous learners, teachers must help students learn to regard errors as an important part of the learning process rather than as a mark of failure. For teachers, mistakes should be viewed as a window for understanding students' misconceptions; for students, errors should be regarded as a tool for reflection and communication about their own thinking. When errors are accepted as part of the learning process, students are willing to tackle more challenging problems, communicate their ideas to other students, and actively reflect on and question the ideas of their peers. This happens in even the youngest grades. In the author's classroom, one preschool child remarked to his best friend, who insisted on counting the dots on each die separately in a board game they were playing, "Why do you do that? It takes too long. You could just *count all* of the dots together."

Mathematics teachers have the difficult challenge of pairing mathematics content with learning procedures that allow students to develop conceptual understanding. This is far more challenging than simply telling students to do x, y, and z. Teachers must create situations in which students encounter mathematical ideas in ways that make sense. That is why incorporating mathematics into daily living situations is such an important part of the mathematics curriculum in preschool and kindergarten. In addition, teachers must introduce materials and activities that challenge children to think mathematically and stay engaged. By building on play—children's natural way of making sense of their world—teachers frame mathematics as the exciting, interesting, and creative process that it is. Through a game-based mathematics curriculum, young children learn to apply the knowledge they already have to new situations and to reflect on the various strategies that their peers use to solve the same problems.

Mathematics Content

Teachers of young children need to understand relevant mathematics content. For teachers of preschool and kindergarten children, this involves foundational concepts in the five content areas of mathematics, as well as an understanding of the content and procedures that children will encounter in the elementary grades. For example, preschool and kindergarten teachers should know and understand categories of geometric forms and the vocabulary that describes them. Otherwise, they cannot guide children in accurately naming and analyzing these forms.

Most important, teachers need to engage in mathematical thinking themselves. In other words, they must recapture the autonomous spirit that they probably had as young children before formal education may have squelched it. How do adults learn to embrace a subject that they may not like? They can start with baby steps. Teachers could try a Sudoku puzzle online or in the newspaper or buy a book of simple math puzzles. At first, old fears and feelings of incompetence may come flooding back. Adults should follow the previously quoted math professor's advice: Try something. Follow some of the suggestions offered online for Sudoku or draw a picture of the math problem in a puzzle book. Share the problem with a friend and see what strategies that person would use. The goal is to reawaken the brain to mathematics.

Any teacher who currently teaches mathematics, or will in the future, is strongly urged to join a professional teachers' organization. This recommendation pertains to virtually all early childhood teachers, including those who intend to teach preschool and kindergarten.

Professional organizations support teachers at all levels through journals, web sites, and conferences. For example, NCTM publishes a monthly journal for early childhood and elementary school teachers called *Teaching Children Mathematics*. The journal includes curriculum articles, many written by teachers; exposure to new research that can inform practice; and mathematics problems to explore with children. Working through problems that may be posed for elementary school children helps preschool and kindergarten teachers focus on their own application of mathematical processes and stay abreast of curriculum that their students will explore in the future.

SUMMARY

Preschool and kindergarten are periods of rapid learning for young children. Because school-entry math skills are predictive of future success in school, the mathematics curriculum in preschool and kindergarten is particularly important. All children bring some informal knowledge of mathematics to their first school experience, although this knowledge can vary widely, depending, in part, on the children's previous experiences. It is important for teachers to connect the informal knowledge of young children to the mathematics experiences they encounter in preschool and kindergarten.

Play is a natural learning forum for young children. Children often incorporate math into their play, such as giving one object to each play partner, figuring out how many items they have when an object is added or taken away, or dividing materials equitably. Teachers can greatly increase the math experiences of young children by joining the play and injecting math-related comments or questions into the dialogue. It is important for teachers to regularly incorporate the language of math into their interactions with young children. If children cannot connect mathematical concepts to related language, learning gaps may occur when they receive more formal educational experiences.

Both small- and large-group experiences can be used successfully with young children if they are carefully designed. Small-group activities, such as interactive math games, allow children to share ideas with their peers and model for one another. Teachers can support children's learning in small-group formats by modeling just above the children's current level of understanding. Large-group experiences allow teachers to connect mathematics to other areas of the curriculum through book sharing, music activities, and counting songs and games. By integrating mathematics throughout the curriculum, as well as designing small- and large-group activities, teachers maximize children's opportunities to construct math concepts.

NCTM has developed standards for mathematics education that are widely accepted and that include goals for preschool and kindergarten. There are five content standards, which encompass the knowledge children should learn: Number and Operations, Algebra, Geometry, Measurement, and Data Analysis and Probability. There are also five process standards, which incorporate the strategies children use to solve mathematical problems and communicate their thinking: Problem Solving, Reasoning and Proof, Communication, Connections, and Representation. Content and process standards are closely connected and are equally important.

Many preschool and kindergarten classrooms now include children with disabilities. Concepts of UDL ensure that all children can participate to the maximum extent possible within traditional classrooms. These concepts include multiple ways for students to acquire and express learning and multiple methods of motivating all students. Accommodations should be made to curriculum activities before they are introduced so that all students can participate.

(continued)

SUMMARY (continued)

Piaget and Vygotsky have greatly influenced the development of instruction for young children. Both emphasized that children construct knowledge through their interactions with objects and people in their environment. Piaget emphasized that mathematical knowledge must be invented by the child through reflecting on experiences. Vygotsky introduced the concept of a ZPD, a zone in which children who are close to understanding a concept can benefit from instruction by more knowledgeable individuals; however, children who are not within this zone will not benefit from the instruction. On the basis of an understanding of these theories, effective teachers provide many experiences in which children can put all kinds of objects into many different relationships. Effective teachers constantly analyze the understanding of their students and provide comments, questions, and modeling that are just above the children's current thinking level.

ON YOUR OWN

- Think of three activities that involved math and that you enjoyed as a child. These activities could be as varied as playing board games, keeping bowling scores, or comparing baseball statistics.

- Try to remember a negative math experience that you had. What could have made this experience more positive?

- Find a math puzzle to share with your friends, fellow students, or colleagues. Compare notes about the ways in which each person approached the puzzle.

REFERENCES

Andrews, A.G., & Trafton, P.R. (2002). *Little kids—powerful problem solvers: Math stories from a kindergarten classroom.* Portsmouth, NH: Heinemann.

Baratta-Lorton, M. (1976). *Mathematics their way.* Menlo Park, CA: Addison-Wesley.

Baroody, A.J., Lai, M., & Mix, K.S. (2006). The development of young children's early number and operation sense and its implications for early childhood education. In B. Spodak & O.N. Saracho (Eds.), *Handbook of research on the education of young children* (pp. 187–221). Mahwah, NJ: Erlbaum.

Boaler, J. 2008. *What's math got to do with it?* New York: Penguin Books.

Clements, D.H., & Sarama, J. (2007). Early mathematics learning. In F.K. Lester, Jr. (Ed.), *Second handbook of research on mathematics teaching and learning* (pp. 461–555). Reston, VA: National Council of Teachers of Mathematics.

Cobb, P. (2007). *Putting philosophy to work: Coping with multiple theoretical perspectives.* In F.K. Lester, Jr. (Ed.), *Second handbook of research on mathematics teaching and learning* (pp. 3–38). Reston, VA: National Council of Teachers of Mathematics.

Copple, C., & Bredekamp, S. (2009). *Developmentally appropriate practice in early childhood programs.* Washington, DC: National Association for the Education of Young Children.

Duncan, G.J., Dowsett, C.J., Claessens, A., Magnuson K., Huston, A.C., Klebanov, P., et al. (2007). School readiness and later achievement. *Developmental Psychology, 43*(6), 1428–1446.

Erikson, E.H. (1950). *Childhood and society.* New York: Norton.

Hatano, G. (2003). Foreword. In A.J. Baroody & A. Dowker (Eds.), *The development of arithmetic concepts and skills: Constructing adaptive expertise.* Mahwah, NJ: Erlbaum.

Hausken, E.G., & Rathbun, A. (2004). *Mathematics instruction in kindergarten: Classroom practices and outcomes*. San Diego: American Educational Research Association.

Kamii, C. (1982). *Number in preschool and kindergarten: Educational implications of Piaget's theory*. Washington, DC: National Association for the Education of Young Children.

Klibanoff, R.S., Huttonlocher, J., Vasilyeva, M., & Hedges, L.V. (2006). Preschool children's mathematical knowledge: The effect of teacher "math talk." *Developmental Psychology, 24*(1), 59–69.

Moomaw, S., & Hieronymus, B. (1995). *More than counting*. St. Paul, MN: Redleaf Press.

Moomaw, S, & Hieronymus, B. (1999). *Much more than counting*. St. Paul, MN: Redleaf Press.

Mooney, C.G. (2000). *Theories of childhood*. St. Paul, MN: Redleaf Press.

National Council of Teachers of Mathematics. (1989). *Curriculum and evaluation standards for school mathematics*. Reston, VA: Author.

National Council of Teachers of Mathematics. (2000). *Principles and standards for school mathematics*. Reston, VA: Author.

National Council of Teachers of Mathematics. (2006). *Curriculum focal points*. Reston, VA: Author.

National Mathematics Advisory Panel. (2008). *Foundations for success: The final report of the National Mathematics Advisory Panel*. Washington, DC: U.S. Department of Education.

National Research Council & C.T. Cross, T.A. Woods, & H. Schweingruber (Eds.). (2009). *Mathematics learning in early childhood: Paths toward excellence and equity*. Washington, DC: The National Academies Press.

National Research Council, Committee on Early Childhood Mathematics (C.T. Cross, T.A. Woods, & H. Schweingruber, Eds.). (2009). *Mathematics learning in early childhood: Paths toward excellence and equity*. Washington, DC: The National Academies Press.

Philipp, R.A. (2007). Mathematics teachers' beliefs and affect. In F.K. Lester, Jr. (Ed.), *Second handbook of research on mathematics teaching and learning* (pp. 257–315). Reston, VA: National Council of Teachers of Mathematics.

Salend, S.J. (2008). *Creating inclusive classrooms: Effective and reflective practices* (6th ed.). Upper Saddle River, NJ: Pearson Merrill Prentice Hall.

Schielack, J.F., & Chancellor, D. (2010). *Mathematics in focus, K–6*. Portsmouth, NH: Heinemann.

Tomlinson, H.B., & Hyson, M. (2009). Developmentally appropriate practice in the preschool years: Ages 3–5. In C. Copple & S. Bredekamp (Eds.), *Developmentally appropriate practice in early childhood programs*. Washington, DC: National Association for the Education of Young Children.

Tudge, J.R., & Doucet, F. (2004). Early mathematical experiences: Observing young Black and White children's everyday activities. *Early Childhood Research Quarterly, 19*, 21–39.

Vygotsky, L.S. (1978). *Mind in society*. Cambridge, MA: Harvard University Press.

Wadsworth, B.J. (1989). *Piaget's theory of cognitive and affective development*. New York: Longman.

2

Developing Number Sense

One, two, buckle my shoe . . .

Five little ducks went out to play . . .

Over in the meadow, near a hole in a tree, lived an old mother owl and her little owls three . . .

Whether in the home or at school, naming and repeating numbers traditionally has been at the core of mathematics education in the early years. This is evident in the rhymes, fingerplays, songs, and counting books that populate early childhood classrooms and that have been passed down to parents through the generations. Research has shown, however, that early mathematics learning in general and the development of number sense in particular involve much more than reciting poems or singing counting songs. Research has unveiled the capacity of children to develop substantial knowledge in the number-sense area prior to first grade (Baroody, Lai, & Mix, 2006; Clements, Sarama, & DiBiase, 2004; Perry & Dockett, 2002). In addition, this early number sense is a strong predictor of later mathematics achievement (Clements & Sarama, 2007). There is, however, a substantial achievement gap between children who live in economically deprived communities and their peers from more affluent environments (Griffin, Case, & Siegler, 1994). Although quality preschool and kindergarten programs are important for all children, research indicates that they make a greater difference for children from low socioeconomic environments (Thomson, Rowe, Underwood, & Peck, 2005).

The ability of preschool and kindergarten teachers to teach mathematics is clearly important. As of 1999, 70% of 4-year-old and 93% of 5-year-old children in the United States attended preprimary educational programs (National Center for Education Statistics, 2001). Given the importance of the development of number sense during the early years, the responsibility for supporting this important learning falls to preschool and kindergarten teachers. They must supply the learning opportunities for children to construct the foundational mathematical knowledge that they will need for their future mathematics learning. For this reason, every possible opportunity must be taken to inject mathematical language and learning into preschool and kindergarten classrooms.

THE NUMBER AND OPERATIONS STANDARD

Number sense encompasses a broad conceptual understanding of number, including the relationship among numbers, the ability to decompose numbers, understanding of the number system, and the relationship among arithmetic operations (NCTM, 2000). This chapter focuses on young children's conceptual understanding of number. Chapter 3 builds on this understanding by exploring children's emerging use of arithmetic operations.

What do mathematics educators mean by a broad conceptual understanding of number? How do children demonstrate this understanding? The three examples that follow illustrate aspects of the development of number concepts in young children of various ages.

EXAMPLE 2.1

Tyler's dad hands him a cracker to keep him occupied while they wait for Dad's coffee order to arrive. "More," demands 2-year-old Tyler. Dad hands him another cracker. Satisfied, Tyler happily nibbles at one cracker from each hand.

Tyler is beginning to understand the concept of *more* when applied to very small quantities. Gradually, with many experiences, this conceptual foundation will expand to encompass the comparison of numerical quantities regardless of size. Tyler also may be exploring the concept of one-to-one correspondence. Toddlers often seem content when they have one object for each hand.

EXAMPLE 2.2

A kindergarten class sings "Five Little Ducks" while the teacher models with toy ducks. After the first verse, one duck does not return. "How many are left?" the teacher asks. The 5-year-olds count the remaining ducks and announce, "Four!"

Across the hall, a first-grade class sings the same song. Their teacher also asks how many ducks are left when the fifth duck does not return. "Four!" the children immediately respond, without bothering to count.

Although both classes came up with the same response, their strategies reflect developmental differences in their understanding of number. The first-grade children immediately determined that four ducks were left. They may have done this in several ways. Some may have visually recognized the remaining ducks as a set of four, without needing to count them. To identify a quantity immediately is called to **subitize;** this idea is discussed further in the next section. Other children may have developed a mental number line on the basis of many experiences with counting objects. These children would reason that four is the number below five, so if the fifth duck is gone, then four must remain. Still other children may have decomposed the five ducks into sets of one and four. All of these problem-solving strategies demonstrate a grasp of number-sense concepts.

The kindergarten children solved in a different way the problem of how many ducks were left. They simply counted the remaining ducks. This also demonstrates number sense, albeit at a less sophisticated level than the first grade children. Whereas the 5-year-olds did not apply the concept of a number line, nor did they decompose a set of five, they did understand that they could determine the answer by counting the remaining ducks. This is **cardinality,** or the understanding that when you count a group of objects, the last number word that you say indicates the total number in the group. Embedded in the children's ability to successfully count the ducks were two other important counting principles: the names and order of the counting words, and the application of one (and only one) counting word to each duck. Understanding of these counting principles is a critical component of the development of number sense.

EXAMPLE 2.3

At group time, a preschool class passes a clear jar from person to person while they chant:
"Who put a cookie in the cookie jar?"
"[Child's name] put a cookie in the cookie jar."
"Oh my, who's next to try?"
After each child has added a cookie to the jar, the teacher asks, "How can we find out how many cookies we have in the jar?"
Several children immediately reply, "We can count them."
"That's one way we could find out," responds the teacher. "Can anyone think of another way we could find out how many cookies are in the jar?"
The class sits quietly for what seems to be a long time. Suddenly, Carla raises her hand. "I know!" she excitedly tells the class. "We can just count the people."
Many of the other children look dubious. "Look," Carla explains. "Each one of us put a cookie in the jar, so if we count the people, then that's how many cookies went into the jar."
The class tries both methods of solving the problem. They discover that they get the same answer by counting the cookies as by counting the people, which was the teacher's intent when he developed the activity and posed the questions.

In this vignette, the teacher pushes children forward in their conceptualization of number by encouraging them to think about numerically equivalent sets. As expected, most of the children determined that to find out how many cookies were in the jar, they needed to count the cookies. Because the activity had focused heavily on the process of each child's adding a cookie to the jar, the teacher expected that some children would be ready to make the logical connection between the quantity of cookies and the quantity of people. He was correct. Carla not only made that connection but was able to explain it to the rest of the class.

THE DEVELOPMENT OF NUMBER SENSE

The examples in the previous section illustrate several ways in which young children demonstrate their understanding of number. How did these children get to this point? Their responses are predictable when educators have an understanding of the developmental trajectories of number-sense concepts in children. More important, this understanding allows teachers to develop a curriculum and spontaneously insert comments and questions that support children's growth along these learning paths.

The development of number sense in young children has been heavily researched during the past three decades, and there is now substantial information to inform instructional practice (Clements & Sarama, 2007; National Research Council [NRC], 2009; Perry & Dockett, 2002). Understanding children's mathematical development in the following three areas of the number-sense domain is critical for teachers of preschool and kindergarten teachers:

1. Quantification

2. Counting

3. Representing number

Quantification

Quantification begins very early in life. Even infants have some implicit knowledge of number, as evidenced by their ability to distinguish between sets of one and two objects, or even two and three objects. This implicit knowledge of quantity is sometimes referred to as subitizing. Researchers have made this determination through experiments that involve habituation. Infants are shown sets of the same quantity of objects (e.g., 2) until they become bored (habituated) and lose interest in looking. The infant is then shown a set of objects that either differs in quantity from the habituated set or is identical, and looking time is measured. If looking time increases when the set size changes, then the researcher assumes that the infant has made a distinction between the sets (NRC, 2009). It is important to acknowledge that animals also demonstrate impressive quantitative abilities (Krasa & Shunkwiler, 2009). Their survival depends on it. As an example, my cat, who lives in our loft room, will not come downstairs until all three of our identical-appearing white dogs are outside. If two dogs go out, but not the third, the cat stays upstairs. If the third dog later goes out, and neither of the other two have come in, the cat quickly comes downstairs for some petting time. He never makes a mathematical error. The dogs are big and like to chase him. Whether human or animal, a rudimentary sense of quantity is apparent from early in life.

Of more importance to teachers is an understanding of how quantification develops. There is a predictable developmental progression. When children are shown a set of objects and asked to produce a set with the same number of objects, they progress through three levels of understanding (Kamii, 1982; Piaget, 1952).

Levels of Quantification

1. *Global quantification*—At this level, children rely on their perceptions to determine quantity. They make a visual or tactile approximation of the set that they are attempting to match. For example, if a child wants to take as many oyster crackers as the child next to her, she may take a handful because it appears to be a similar amount. If the teacher shows her a row of pennies, she may make a row that is about the same length as the original row without regard to the actual number of pennies in each row.

2. *One-to-one correspondence*—Children at this stage select one item in their new set for each item in the existing set. They may carefully align the objects or glance back and forth from the original set to their own set each time they take an object. This stage shows an important development in children's quantification because they now are able to focus on the units included in the set rather than just its global parameters.

3. *Counting*—Children at this stage count the items in the original set and then count an equivalent number of objects for their matching set. In order to use counting as a strategy for quantifying, children must understand that the last number they count represents the total.

Attaching ages to these stages is unreliable because different children progress at very different rates, depending in large part on their individual experiences. What is important is for teachers to understand the developmental progression so that they can model a quantification level at or slightly above the child's current level of thinking. Recall that there is strong evidence that even infants can subitize small sets of one, two, and three objects. Therefore, preschool teachers should expect 2- and 3-year olds to subitize small sets. Even toddlers will want two crackers rather than one! Preschool children who see two dots on a card reliably take two counters to represent the set (Moomaw, 2008), and most will correctly identify the number as two.

For quantities larger than two or three, preschool teachers are likely to see both global and one-to-one correspondence responses from children attempting to create equivalent sets, especially among the younger children. It is surprising that, even though many preschool children can count objects, they often do not use counting to compare the quantities of objects in collections (Clements & Sarama, 2007). However, teachers who use a game-based curriculum, as is advocated in this text, often find that many preschool children are at the counting level of quantification. In one study, 62% of a sample of 107 children 3–5 years of age who attended a preschool that used a game-based mathematics curriculum (Moomaw & Hieronymus, 1995, 1999) accurately used counting to compare sets of three or fewer items on a curriculum-based mathematics measure, and 41% used counting to compare sets of four or five items (Moomaw, 2008). Kindergarten teachers should also expect to see a wide range of quantification levels among the children they teach, from children still using one-to-one correspondence to compare sets, to children combining sets and comparing the results.

Counting

Accurate, meaningful counting is an essential early mathematics tool that predicts later success in elementary school arithmetic (Aunola, Lesinen, Lerkkanen, & Nurmi, 2004). Young children love to count and eagerly participate in counting games, songs, and rhymes. Meaningful counting, however, is much more complex than simply saying a string of counting words. In order to accurately use counting to quantify, make comparisons, and solve problems, children must understand five principles of counting (Gelman & Gallistel, 1978). The first three of these, which are sometimes referred to as the "how to count" principles, are likely to develop during the preschool years; the fourth and fifth principles involve more extended reasoning about counting and its applications. These latter two take longer to develop.

Principles of Counting

1. *Stable order*—In order to successfully count a group of objects, children must apply the designated counting words adopted by their culture in a correct, stable order.

2. *One-to-one correspondence*—Children must understand that each item in a group is counted one, and only one, time.

3. *Cardinality*—To use counting in a quantitative sense, children must realize that the number word given to the last item counted in a group represents the total quantity.

4. *Order irrelevant*—Children who fully comprehend the meaning of counting realize that the order in which the objects are counted does not affect the total. They understand that the number words are applied only temporarily to the objects being counted and have nothing to do with the objects themselves.

5. *Abstraction*—Eventually, children comprehend that counting can be applied to the quantification of both concrete and abstract entities; for example, ideas can be counted.

Teachers who provide many opportunities for children to quantify and think mathematically, especially during play, will observe children regularly applying aspects of the counting principles. Each episode provides assessment data to the teacher, who may respond with questions or comments that extend the children's mathematical thinking or may use the information for future interactions or planning.

Children's construction of the first three principles, the "how to count" principles, does not occur in a specific order, nor does it happen all at once. For example, many young children apply the one-to-one correspondence rule when counting the first few objects in a set, but then begin skipping some objects as their pointing finger moves faster than they say the number words. Conversely, children may appear to forget which objects they have already counted and count them more than once. In terms of stable order, many children know the order of the counting words up to a certain point, but then say numbers in an incorrect order or repeat numbers that they have already used. It is interesting, though, that even after they have used their memorized string of counting words, children usually fill in with number words rather than with other labels, such as letter names. Finally, some children can count objects but do not understand that the last object they counted represents the total for the entire group. They have not yet constructed cardinality. The assessment section of this chapter focuses on how to analyze children's counting in a variety of situations and discusses how this information guides teachers' instructional practice.

Application of the order-irrelevant principle is rarely apparent in preschool, but may begin to emerge during kindergarten. The following is an example of one child's "ah-ha" moment when she suddenly made the connection between the order-irrelevant principle and cardinality.

EXAMPLE 2.4

The students in a summer kindergarten class were interested in how many days they would meet over the summer. Each day, during the beginning circle time, the teacher added one teddy bear shape to the bulletin board to represent that day. Together, the teacher and children then counted all of the bears to determine how many days they had already met. They always counted the row of bears from left to right, in typical reading fashion.

One day the teacher decided to try an experiment. When it was time to count the bears, she counted from right to left rather than from left to right. The children protested vociferously. "Now we have no idea how many days of summer school we've had so far," they informed her. Surprised, the teacher asked, "Do you want me to count in the other direction?" The reply was a definitive and unanimous "Yes!" The class then counted the bears from left to right. None of the children seemed to notice that the total was the same as the total the teacher had previously counted.

Intrigued by their response, the teacher decided to repeat the procedure on the following day. Once again the children protested. The bears were re-counted in the "correct" direction, and the children seemed oblivious to the fact that the totals from both directions were the same. The same scenario unfolded each day for two weeks. On the 10th day of the teacher's experiment, something different happened. Once again, the teacher had counted from right to left, the children had protested, and the teacher had asked whether she should count from left to right. Suddenly, Angela began bouncing up and down on her mat in excitement. "Wait, wait, wait!" she exclaimed. "It doesn't matter. Whichever way you count, you still get the same answer." This was a big moment for Angela and a revealing one for the teacher. Angela had just figured out the principle of order irrelevance and accurately explained it to her classmates. Her teacher was excited to have witnessed this event. She realized that her "off the cuff" experiment had provided the experiences that Angela (and, shortly thereafter, many of her classmates) had needed to understand this big idea. The teacher integrated this knowledge into her pedagogy and began counting sets of objects in various orders during many kindergarten math experiences.

What about the abstraction principle? Would teachers ever see an example of children wrestling with that idea during preschool or kindergarten? Actually, an example earlier in this chapter described a group of children who took turns, each adding a cookie to a jar. The children were confident that they could find out how many cookies were in the jar by simply counting the cookies; however, they were stumped when their teacher asked for another way to find out how many cookies they had. After some concentrated thinking, one child realized that they could count the people instead of the cookies. This is a good example of abstraction. Rather than directly count the objects (cookies), the child realized that she could, instead, count the people who put the cookies into the jar. In essence, she connected counting to a sequence of events, the placing of the cookies into the jar.

In each of the last two examples, one child constructed an important mathematical concept and reported it to the group. A teacher might wonder about the rest of the children. Are they benefiting from these activities? As noted throughout this book, the dialogue that emerges as children share mathematical discoveries is an important component of the development of the entire group. As children attempt to understand an idea coming from one of their peers, they think harder about what this means. Piaget would say that the children are in **disequilibrium** as they try to reconcile a point of view that differs from their own (Wadsworth, 1989). If the idea is close to what they are ready to understand, or within what Vygotksy (1978) termed the ZPD, then learning will likely occur. The ideas that peers share generally are close to what others are ready to understand, because they are near the same developmental levels. This is why mathematical discourse is such an important component of the curriculum.

Representing Number

Yet another important component of number sense is children's ability to represent number. Young children do this in many ways. Acting out problems, drawing, modeling with materials that can be manipulated, gesturing, and applying conventional numerals are all ways in which young children represent number. Representation of numerical relationships often occurs through children's play. Table 2.1 illustrates typical ways in which young children represent number concepts.

Symbol Systems

Piaget distinguished between two kinds of symbol systems: 1) a **symbol,** which looks similar to what it represents, and 2) a **sign,** which is an arbitrary social construct that does not look similar to what it represents. This is an important distinction for children's understanding of mathematics (Kamii, 2000). Symbols are, in a sense, pictorial representations that can be invented by the child. Thus, a child who is asked to represent three apples might draw

Table 2.1. Children's representation of number concepts

Experience	Number concept
A toddler lines up his dolls and stuffed animals and gives each one a pretend sip from a cup.	One-to-one correspondence; each toy gets one sip
A child draws a picture of her family. There are large figures for Mommy and Daddy and smaller figures for the child and her sister.	Quantification; the child represents the four people in her family through drawing
Three children stand on a pile of mats in the gross motor room. Together they count, "1, 2, 3, 4, 5." Then they all jump off.	Rote counting; children count the first five numbers in order
A kindergarten child makes a birthday card for his friend. He writes a big numeral "6" on the card because she will be 6 years old.	Numeric representation
A child gets a "fun meal" at a restaurant. His dad asks him how many toys are in the box. The child sees two cars and holds up two fingers.	Representation of number through gestures
A child rolls a die and sees three dots. She takes three toy frogs from a bowl and sets them in front of her.	Quantification; creation and comparison of sets represented through manipulative materials

three circles or make three hash marks. These symbols help the child denote the units that are represented. Numerals, on the other hand, are considered signs. They do not resemble what they represent and cannot be invented by the child; instead, the number word that corresponds to each numeral must be directly taught.

Many children who enter kindergarten are able to name numerals. They may even be able to use their fingers or other objects to show the quantity represented by the numeral. Their ability to use numerals in a mathematical sense, however, often is limited. Recent research indicates that kindergarten children do not select numerals as their way to represent quantities; instead, they use pictographic representations (Kato, Kamii, Ozai, & Nagahiro, 2002). Teachers of young children who insist on the use of numerals may hamper children's ability to comprehend the mathematical concepts that the teachers are trying to teach.

DESIGNING THE NUMBER SENSE CURRICULUM

Mathematics should be as interwoven into the fabric of the preschool and kindergarten curriculum as literacy currently is. During the first decade of the twenty-first century, attention focused heavily on the early literacy curriculum. Teachers of young children learned to design literacy activities for individual students, and literacy became the focus of regular small- and large-group experiences. In addition, many teachers began successfully to integrate literacy activities throughout the classroom. The same emphasis must be placed on the integration of mathematics into the curriculum. Children in the United States continue to trail their peers in Asian and European countries in mathematics, as reported by the National Center for Education Statistics (NCES) in the Trends in International Mathematics and Science Study (NCES, 2004).

By weaving mathematics throughout the curriculum, teachers help children understand the connections between mathematics and the real world. They can focus on individual children, build upon peer interactions, and draw the group into mathematical discussions. For these reasons, curriculum planning is divided into four broad areas:

1. Math talk

2. Individual or small-group activities

3. Large-group activities

4. Number sense throughout the curriculum

Math Talk

Math conversations in the number-sense area often involve creating one-to-one correspondence relationships, creating and comparing sets, and counting, because these are foundational concepts. Initial experiences with pairing objects from two groups in a one-to-one fashion provide a foundation for later pairing one number word with each object counted. For this reason, teachers of young preschool children may focus on the concept of one-to-one correspondence in their interactions and conversations with them. For example, in the manipulative or block area, they might draw attention to small animals. The teacher might say, "Look at all these animals. Can you help me find a block for each animal to stand on?" This introduces the possibility of a one-to-one relationship between animals and blocks. Following are some more examples of teacher comments or questions that can lead children to think about one-to-one correspondence:

•   Dramatic Play—*Kim says that some friends are coming for dinner. Can you find an egg for each one of these plates?*

•   Water Table—*Do we have a funnel for each bottle, or do I need to find some more?*

- Art Area—*I'm getting out the paint. Can you find a brush for each paint jar?*

- Music Area—*I'm going to play each chime one time and see how it sounds. Do you want to try it?*

- Gross Motor Area—*Let's play a game in which we hop one time on each carpet square.*

Both preschool and kindergarten teachers should place a major focus on affording natural opportunities for children to create and compare sets and count. At first, these experiences should involve small quantities; however, larger amounts to quantify and compare can be gradually introduced. The following are examples of comments and questions that focus children's attention on these concepts:

- Manipulative Area—*Look at all the grasshoppers in this collection of insects. How many do you think there are? Can you help me find out if there are as many ladybugs as grasshoppers in the basket?*

- Sensory Table—*We have oats in the sensory table today. Did you know that horses eat oats? That's why I put horses in the table, too. Here's a bucket for this horse's oats. Let's count how many scoops will fit into his bucket.*

- Book Area—*Look at all the animals in this farm picture. I wonder if there are more sheep or cows. How about this—I'll count the sheep and you count the cows, and we'll compare.*

- Art Area—*Look at all the flowers on your picture. Did you put the same number of petals on each flower, or a different number?*

- Snack—*I'm going to use the apple cutter to slice this apple. How many pieces did it make? Are there enough apple slices for everyone at the table to have one? Can everyone have two slices?*

The sections that follow discuss developing number sense curricula for individuals or small groups, large groups, and for integration into other areas of the curriculum. In each case, the math talk that surrounds the activity is vital. The activities give teachers the opportunity for an ongoing assessment of quantification levels, counting principles, and representation. On the basis of their observations and interactions, teachers can respond with comments and questions that support each child's mathematical growth.

Individual or Small-Group Activities

Small-group math activities are highly recommended for helping children develop mathematical thinking. In small-group situations, children can interact with one another and more easily focus on the activity. The teacher can monitor these small groups, assisting when necessary and letting children resolve differences when they are able to. Interaction among peers should be encouraged. Children can check one another's responses for errors, which fosters discussion and mathematical thinking. Peers may also model a higher level of quantification than their play partners. This may accelerate development, as children at less advanced levels of reasoning try to figure out what their friend is doing.

Researcher Constance Kamii has long advocated a game-based curriculum for mathematics education in preschool, kindergarten, and the primary grades (Kamii, 1982, 1994, 2000, 2004). Building on her ideas, teachers at the Arlitt Child and Family Research and Education Center at the University of Cincinnati developed an extensive array of math games and activities for preschool and kindergarten children (Moomaw & Hieronymus, 1995, 1999). Recent research has supported the use of math games for numeracy development in young children (Ramani & Siegler, 2008; Young-Loveridge, 2004; Whyte & Bull, 2008).

Four types of number sense games are presented in this section: math manipulative games, grid games, path games, and card games. They can be played by individual children, a child and a teacher, or by pairs or small groups of children. The games accommodate children at all three levels of quantification, encourage understanding of the counting principles, and involve representation of number. They therefore strongly support children's development of number sense. An example of each type of game is provided, along with suggestions for implementation.

Math Manipulative Games

Math manipulative games have two components: 1) a die, spinner, or deck of cards with dots or pictures for children to quantify; and 2) a set of objects to collect. Players take turns rolling the die, spinning the spinner, or drawing a card. They quantify the amount on the card and then take a corresponding number of counters. Math concepts embedded in these games include quantification strategies and creation and comparison of sets.

Symbols (not numerals) are used on the die, spinner, or cards for several reasons:

1. Symbols allow children at all three levels of quantification (global, one-to-one, and counting) to use the game, and they encourage progression to the next stage of development. If numerals are used, the game is limited to children at the counting stage who also recognize numerals.

2. Symbols, such as dots or stickers, present numeric representations in both standard and nonstandard configurations. This is important because, with repeated experiences, children begin to realize that the positioning of the dots or counters does not change the quantity that is represented.

3. Symbols support composition and decomposition of numbers because children can see the units. This allows for a natural progression into addition and subtraction when children are ready (see Chapter 3). Numerals require children to mentally picture the units, which many cannot yet do.

All sorts of counters can be used for math manipulative games: sea shells, small pine cones, cloth or plastic fruits or vegetables, small toy animals, pompoms, buttons, and so forth. Party supply stores and dollar stores are good sources of interesting and inexpensive counters. Of course, teachers must always select a size and type of counter that is safe and appropriate for their class.

ACTIVITY 2.1

The Bathtub Game

Materials

The following materials are needed for this game:

- ▪ 1–3 dot die, made from a 1-inch cube and ¼-inch file folder stickers (two sides have 1 dot; two sides have 2 dots; two sides have 3 dots)

- ▪ Two plastic bathtub soap dishes or oval-shaped bowls

- ▪ One basket holding approximately 20 toy people (e.g., Duplo, Playmobil, Fisher Price figures)

Description

This game was created to coordinate with the popular children's book *King Bidgood's in the Bathtub,* by Audrey Wood (2005). In the story, the king refuses to get out of the bathtub, so all court functions must take place in the tub. More and more people join the king in the bathtub. Children find especially appealing math games that coordinate with books they love. Therefore, this simple game gets a lot of use.

To play the game, children take turns rolling the die and placing a corresponding number of toy people into their bathtub. Children at the global stage of quantification may randomly take some toy people and put them into the tub. Children at the one-to-one correspondence stage may point to the dots one at a time and place a toy person into the tub each time they touch a dot. Children who can count may use that strategy. Because the numbers on the die are small, some children may subitize. After all of the toy people have been used, children may wish to compare how many toy people they each have in their respective tubs. They may align them in rows to compare quantities, or, perhaps, count them.

If this game is too easy for some children or for the class as a whole, teachers may change to a standard 1–6 dot die and use smaller toy people or even teddy-bear or frog counters. More of these will fit into the bathtubs. The beauty of teacher-made math games is that they can be modified to fit the needs of specific children or classes. They also can be coordinated with popular topics of interest. Each time children take a turn, they have a new math problem to solve. The games provide repeated opportunities to quantify and represent number within a play situation.

Math Discussions

The conversations that occur while children are playing math games are an important part of the learning. They familiarize children with the language of math and stimulate thinking. Teachers must learn to ask relevant, interesting questions that are related to the game, but target each child's developmental level. Comparison questions are particularly useful. The following are some examples of comments and questions that teachers might use with the bathtub game:

- *Which bathtub do you think has more people so far?*
- *If Crystal rolls a 2, will she have as many people in her bathtub as Kyla?*
- *I had two people in my bathtub, and I rolled another 2. Let me put two people in, and then I want to see how many I have all together.*
- *If all the people wearing blue go in one bathtub, and all the people wearing red go in another bathtub, I wonder which bathtub will have the most people.*

Math Grid Games

Math grid games are similar to math manipulative games, but have an added component: a bingo-type board for placement of the counters. The boards may be divided into boxes (grids) or have rows of small pictures. In addition to taking a quantity of counters that matches the number of items on the die, spinner, or cards, children also place each counter on a space on their game board. This creates opportunities for additional practice with one-to-one correspondence (one counter for each space) and allows children to envision the composition of sets. For example, if a child has a row of five boxes on her board and three of the boxes are filled, she can see that two boxes are left. Teachers can draw attention to the composition of sets through game-related questions such as this: "You have markers on three boxes in this row. How many boxes don't have a marker?"

Doghouse Game

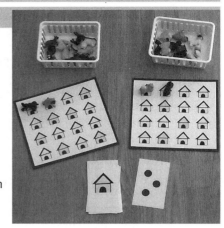

Materials

The following materials are needed for this game:

- Game board for each player, made from poster or tag board (8 × 8 inches)
- Sixteen doghouse pictures for each board, attached to the board in 4 rows, with 4 dog-houses per row
- Sixteen small toy dogs for each player
- Deck of cards, made from 3 × 5 inch file cards, with one to four 1-inch circle stickers per card

Description

This game appeals to many children because they are interested in dogs or other pets. It accommodates children from a wide range of developmental levels. Young, preschool children who are not yet ready to draw cards and create matching sets may be interested in just placing a dog on each doghouse picture. This reinforces the important concept of one-to-one correspondence. If children place the dogs randomly on the board, the teacher can direct their attention to a one-to-one relationship by asking whether the children can find one dog for each house. Many children will take turns drawing the cards to determine how many dogs to place on their boards. Children at the one-to-one correspondence level of quantification may actually set the dogs on the dots on the cards before moving them to the doghouses. Children who can count will use that strategy to determine how many dogs to take.

Grid games can be played by an individual child, and they make excellent small-group games as well. In preschool classrooms, a grid game can be used as a special activity or placed on a math game table. Kindergarten teachers might decide to divide the class into small groups to play the game. As with other grid games, the doghouse game can be modified for children who are more advanced. They might be asked to put two or more dogs in each doghouse. The quantities of dots used on the cards can be increased as well.

Teachers often have to spend their own money for classroom materials, so cost is an important consideration. The doghouse game was selected because dog and cat counters can be purchased in large quantities for relatively little expense. Doghouse pictures can be downloaded from free clip-art sites.

Math Discussions

Questions similar to those used for math manipulative games are also appropriate for grid games. In addition, teachers can draw attention to the number of spaces covered, the number of empty spaces, or the relationship between the two. Note that all of the sample questions and comments listed next relate directly to the game; therefore, although they are excellent math questions, they fit naturally into the flow of the play. It is this natural seeking of information that makes mathematics creative and exciting, not constant quizzing that is unrelated to anything of interest to the child.

- *How many more dogs do you need to fill up all of your doghouses?*
- *One of my rows is full, one row has two dogs, and this row is empty.*

- *Do you have more occupied doghouses or more empty doghouses in this row?*
- *I got only a 1. Now I'll have to roll again to fill up my last two doghouses.*
- *How many dogs do we have houses for?*

Path Games

Path games are more difficult for young children than grid games; however, they are valuable teaching tools because they simulate a number line. A number line is actually a mental construct that individuals create as they learn to rank numbers in a hierarchy. Mathematics curricular materials that may be used in kindergarten and first grade often create a physical number line, with numerals listed in order, for children to use when answering questions about the magnitude of numbers or when solving arithmetic problems. Difficulties occur when children have not yet constructed a mental number line. They may model use of a physical number line without any understanding of what they are actually doing mathematically. When children play path games and move an object along a series of marked spaces to reach their destination, these games model continuous movement along a number line. With repeated experience, children realize that as they move forward, they take an increasing number of steps along this line, and the spaces which they cover can be quantified.

Path games are more difficult for young children than grid and counter games because they are more abstract. Children do not have a tangible object to represent each quantity; instead, they must quantify steps taken toward a goal. As in Hansel and Gretel's journey through the forest, however, the steps they have taken seem to vanish as they move forward. Assume that a child has moved three spaces on his first turn, and then moves two more spaces on his second turn. His place on the path does not look like two; instead, it looks like five—or just "a lot." At first, some children move their marker back to the starting point for each turn so that their placement on the path matches the number on their current turn. Quantifying spaces involves a new entity for children to count, so path games support development of the abstraction counting principle.

A common error that children make when playing path games is to re-count the space which their mover is on when they begin a new turn. They often make the same mistake when they perform addition on a number line (see Chapter 3). Gross motor path games, in which children are themselves the movers, help the children understand that the space they occupy at the end of a turn is not part of their next turn. They realize that it has already been counted because they are standing on it. Many children stop making this re-counting error on board games after they have had experience with gross motor path games.

Games with short, straight paths help children make the transition from concrete manipulative and grid games to more abstract path games. For these initial path games, children should have their own paths to move along so that they are not confused by the markers of other players. Ten spaces is a good length for initial paths. This relatively short length seems to help children envision their movement along the path. If they have moved five spaces, they can see that they have five more to go. Once children have constructed the concept of moving along a path on consecutive turns, they can tackle games with longer paths that have curves or angles, additional directions, and bonus or trap spaces. Children can now share the same path, which allows them to account for the position of their mover in relationship to those of the other players. They can also use dice with larger amounts, and eventually begin to add the numbers on two dice. This transition into addition will be the focus of Chapter 3. The following examples illustrate both short- and long-path games.

ACTIVITY 2.3

Get the Dogs Home— Short-Path Game

Materials

The following materials are needed for this game:

- Game board for each player, made from poster or tag board (6 × 18 inches)
- Ten 1-inch stickers (circles or paw prints) for each game board
- Enough doghouse pictures for the end point of each path to have one
- Small plastic dog for each player
- 1–3 dot die

Description

This game is similar to the doghouse grid game, but in this game players move a dog along a path to reach the doghouse. Small groups of children, each with an individual game board, can play together, or a child can play alone. At first, teachers may notice children hopping their dogs all the way to the doghouse on their first turn. It is helpful for the teacher to model taking one step for each dot on the die while also explaining what she is doing. For example, the teacher might say, "Look. I rolled three dots. I get to move one space for this dot, one space for the next dot, and one space for the last dot." A small number of dots on the die is used to help children make the transition from counting objects to counting steps along the path. Once children have made this transition, a 1–6 dot die can be substituted and additional doggie movers supplied to make the game last longer.

Math Discussions

A new type of discussion can occur as children play transitional path games. Comments and questions can now center around the quantification of spaces that have been accounted for (stepped on) versus spaces that have not yet been counted. This draws attention to the more abstract entity that is now being quantified. The comments that teachers make during the course of the game, as well as their conversational questions, can help children recognize the relationship between the number of spaces that they move on each turn and their total distance along the path. This connects with later work related not only to the use of a number line, but also to measurement. Following are some comments and questions that focus on these relationships:

- *How far has your doggie moved so far? Mine has moved only 2 spaces.*
- *Oh, look. My doggie has moved 5 spaces, and he has 1, 2, 3, 4, 5 more spaces to go.*
- *How many spaces are left before your dog gets to his house?*
- *Our dogs both got home. Let's count how many spaces they had to step on to get to their houses.*

ACTIVITY 2.4

Get the Dogs Home— Long Path Game

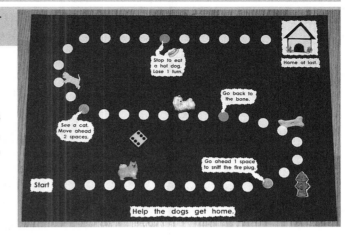

Materials

The following materials are needed for this game:

- Poster board, 22 × 22 inches
- 35–40 1-inch circle stickers, arranged in a large S shape, to create the path
- One doghouse picture, for the end point of the path
- Enough small plastic dog movers for each player to have one
- Four "announcement" stickers, for the trap and bonus spaces
- Bone and flower stickers, for decoration
- 1–6 dot die

Description

The dog theme continues for the long-path game in order to help teachers realize that they can construct a collection of games which accommodate children at three developmental levels: 1) those who need concrete objects to quantify (grid games); 2) those who are in a transitional phase between counting concrete objects and more abstract entities, such as path spaces (short-path games); and 3) those who are comfortable counting along a path (or number line) and can now apply that concept to a larger number of spaces (long-path games). This particular game has clear start and end points. Children progress along the path and try to get their dog movers to the doghouse at the end. Along the way, there are bonus spaces that let them move forward and trap spaces that require them to move backward or lose a turn. These can be eliminated from the game if children do not like them or if they cause confusion.

Teachers can use the photograph as a model for how to construct this game, or experiment with their own design. To eliminate confusion, the stickers should be spaced about ½ inch apart. If the stickers are too close together, children are more likely to skip over them; if they are too far apart, children lose track of the configuration of the path. Along the path, two stickers colored green (for the bonus spaces) and two stickers colored red (for the trap spaces) should appear. Next to these spaces, an announcement sticker with a message should be attached to the board. Some examples of messages are "Move ahead 1 space," "Move back 1 space," "Lose 1 turn," "Go back to the bone," and "Move ahead to the flower."

Math Discussions

This type of game allows for additional mathematical questions that focus on relative positions and probability, as the following examples illustrate:

- *Which dog do you think is going to reach the doghouse first? Why?*
- *I hope I don't get a 3. Then I'll have to move back to the bone!*
- *Which dog is the farthest away from the doghouse?*
- *My dog is in last place. Is there any chance he could get to the doghouse first? What if I roll a 6 on my next two turns?*

Card Games

Various types of card games provide the opportunity for children to quantify and compare sets. Kamii has documented many games that kindergarten children can play to strengthen their awareness of the magnitude of numbers (see Kamii [2004], *Young Children Reinvent Arithmetic*). This section focuses on two card games that can be played with the same teacher-made deck of cards. The first is a variation of the traditional game War, which some mathematics curriculum books now call Top It, a more prosocial name. In this type of game, children draw cards and compare the sets on them. This gives children repeated opportunities to think about more, less, and the same. The second game is similar to the traditional Go Fish game, except that children attempt to match cards with the same number of pictures rather than matching colors, as in some preschool versions of the game.

In these card games, both the objects and the placement patterns on the cards may vary. This variation forces children to consider that the same number can represent sets with different objects as long as the number of objects is the same. For example, a card with 4 stars can be matched to a card with 4 pigs because they both have 4 objects. This relates to the abstraction principle. In addition, children must consider quantities that may be the same, yet look different because the placement of the objects is different. For example, a card with three rows, each with two objects, would match a card with five objects in one row and one object in a second row. This concept relates to the order irrelevant principle because the positioning of the objects does not affect the quantity. Both types of games broaden children's conceptions about the meaning of number.

ACTIVITY 2.5

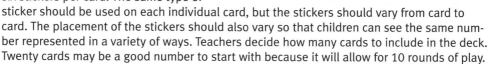

"I Have More" Card Game

Materials

The following materials are needed for this game:

- Assorted stickers
- Package of 3 × 5 index cards, preferably unlined

Description

Create the card deck by applying one to six stickers per card. The same type of sticker should be used on each individual card, but the stickers should vary from card to card. The placement of the stickers should also vary so that children can see the same number represented in a variety of ways. Teachers decide how many cards to include in the deck. Twenty cards may be a good number to start with because it will allow for 10 rounds of play.

To play the game, children first divide the cards evenly between two players. (This in itself is an interesting math activity.) After assembling their cards in piles, face down, children turn the top card face up. The child whose card has the larger quantity gets to take both cards. If both cards have the same amount, the children each turn over another card. The player whose second card has the larger amount takes all four cards. Play continues in this manner until all of the cards are gone. At this point, children compare their total number of cards to determine who has ended up with more.

Even though this game deals with quantities that go up to only 6, it is still not easy for young children. Comparing sets is often more difficult than just counting an individual set. For this reason, the game is usually more appropriate for older preschool and kindergarten children. Not only may younger children have difficulty with the mathematical concepts, they may not want to relinquish their cards to another player! Over time, as children become more advanced, the card deck can be expanded to include larger quantities on each card. Individual children may want to assemble the cards in hierarchical order. This helps them develop an internal number line.

Math Discussions

Dialogue may quickly develop with this game as children try to decide who has more stickers on their card. After all, some of the stickers may be larger than others or may appear to take up more room on the card, but there may be fewer of these than the smaller stickers on another player's card. The teacher's role in these discussions is to mediate, not to provide answers. She may ask children why they think their card has more and help them express their thoughts. Arguments among children should be viewed positively as they lead to advanced mathematical thinking. Following are some examples of teachers' comments or questions that could be used to clarify and resolve disputes.

- *Annabel says her card has more. The pictures take up a lot of room. Why do you think your card has more?*
- *How can we find out for sure which card has more?*
- *Dimitri has been watching. Let's hear what he thinks.*
- *Joe thinks you counted that picture more than once. How can we check to make sure that doesn't happen?*

ACTIVITY 2.6

Matchmaker Game

Description

This game uses the same card deck as the game I Have More. Each child starts with five cards. The children look at their cards and make matches on the basis of quantity. As in the game Go Fish, children then take turns asking each other for cards and attempting to make matches. If the person asked does not have the requested card, then the child who asked draws a card from the deck. Each time a match is made, the child places the matched cards down, face up. This allows other children to check for accuracy.

Math Discussions

This game may create some heated discussions. First of all, some children will not respond appropriately when they are asked for a particular card. This may be because they do not recognize the card as having the requested number designation. Children may also make errors as they attempt to create pairs. The teacher can help by suggesting a checking routine, such as the following:

Astrid, Julie asked you for a 4. Let's look at your first card. Does that have four things on it? Now let's look at the next card.

This procedure helps children become more organized in their problem solving.

Large-Group Activities

Whole-group activities can, and should, be used to support children's mathematical learning. Older preschool and kindergarten children become increasingly interested in the thoughts and responses of their friends, and large-group activities foster the feeling of a mathematics community. A wide variety of mathematics activities easily can be integrated into traditional circle times in preschool and kindergarten. How unfortunate, then, that so many teachers do not take advantage of this opportunity, or limit group-time mathematics to the same daily calendar activities. Research indicates that children do not understand what calendars represent until the third grade or later. This should not be particularly surprising to teachers of young children, who often notice these children's confusion about terms such as *yesterday, tomorrow,* and *next month*. Time would be much better spent helping children quantify things that they can comprehend and representing mathematical ideas in a wide variety of ways. The following anecdote illustrates how one class explored mathematics as part of a whole-class activity.

EXAMPLE 2.5

Mark is the teacher of a class whose members had become very interested in the book King Bidgood's in the Bathtub, *by Audrey Wood (2005). They had shared the big-book version at group time, requested repeated readings of the story throughout the week, written stories about the king, and played at the sensory table with toys similar to those in the book. Mark decided that it would be interesting for the children to act out the story. The children decided to build a large bathtub out of hollow wooden blocks. Each of the 18 children added a block to form the perimeter of the tub. As Mark read the story, designated children played the roles of the various characters and joined the king in the bathtub. When the story was finished, the remaining children also wanted to get into the tub. This was a problem, however, as the bathtub was already full.*

"What should we do?" asked Mark.

"We should make it bigger," responded the children.

"How much bigger do we have to make it?" Mark asked.

The children were not sure. "How many people are already in the tub?" Mark asked. The children counted 10 people.

"How many people are not in the tub yet?" Mark then asked. The children counted 8 people.

"Are there more people inside the tub or outside the tub?" Mark asked.

Anthony said that 10 was more than 8, so there must be more people inside the tub. "Does everyone agree?" Mark asked.

Allison disagreed. "Look at the people," she said. "There are all these people out here. It's more than in there." The children who were not in the tub were sitting in a circle that stretched farther than the length of the tub.

"Hmm," said Mark. "Is there another way we can find out?"

"I know," said Hakim. "Make a row of the people that are in the bathtub, and next to that make a row of the other people." The children in the bathtub climbed out and made a row. Then the remaining children made a row opposite them.

"See," said Hakim. "There are 2 more people in this row."

"I knew it," said Anthony.

"Allison, what do you think?" asked Mark. "Were there more children inside the tub or outside?"

"Inside," agreed Allison. "I think they were just all scrunched up inside the tub."

"That's probably true," agreed Mark. "That can make it hard to tell which is more."

The class decided that each person should bring a second block to extend the size of the bathtub. Soon the entire class, plus Mark, was sitting in the tub.

This math problem involved all of the children in the class. The children's responses represented the three levels of quantification previously discussed. Anthony applied counting to compare the numbers of people inside and outside of the bathtub, but Allison was confounded by her global perspective. No doubt, other children in the class shared her evaluation of the situation. Hakim applied a one-to-one correspondence strategy to solve the problem. This was close enough to Allison's level of reasoning to convince her of her perceptual error. It also confirmed to Anthony that his counting comparison was correct. All children can benefit from these types of group discussions and activities. They discover that people have different opinions and that there are various ways to solve a problem. The teacher's awareness of these thinking levels, as well as of the importance of sharing information, led to his series of questions that deepened the understanding of many in the class.

There are many ways to regularly incorporate number-sense activities into group times. In an earlier example in this chapter, a teacher modeled the mathematical changes in the "Five Little Ducks" song while children sang. Counting songs, books, and poems are an excellent way to support children's development of stable order counting, one-to-one correspondence, and cardinality. By modeling the mathematics, or letting children model the problem, teachers can make the math content more visible. In the activity described next, photographs of the children in the class are added one at a time to a school bus poster while the children sing an adapted version of the familiar song "I Looked in the Window."

ACTIVITY 2.7

The School Bus Song

Materials

The following materials are needed for this game:

- A large outline of a school bus, drawn and cut from yellow poster board
- Small individual photographs of each of the children in the class
- Two strips of magnetic tape, placed in rows along the school bus
- Index cards, each marked with one numeral, from 1 to the total number of children in the class

Description

Many children ride a bus to school, or have siblings or friends who do. This activity, therefore, relates to a familiar experience. Because photographs of the children themselves are used, classes are usually quite interested in this activity. The photos can be laminated for durability and a small piece of magnetic tape attached to their backs. They then can be hung along the magnetic strips on the bus. Either the teacher or a child can add one photograph for each verse of the song. A numeral card can also be held up to represent the number of children on the bus. Children can participate in the song during group time and later repeat the activity during choice time.

> *I looked in the school bus, and what did I see?*
> *I saw 1 child, looking back at me.*
> *I looked in the school bus, and what did I see?*
> *I saw 2 children, looking back at me.*

Continue until all children have been added to the bus.

Math Discussions

This activity reinforces the key concept of hierarchical inclusion: When numbers are counted forward, each successive number includes all of the previous numbers counted. The activity, therefore, illustrates the counting principle of cardinality. Whereas young children may be able to subitize the first three photos that are added to the bus, for successive numbers, many will need to go back and re-count the entire set each time a new photo is added. Teachers should let them do this, as it is a necessary step in their development. Eventually, children realize that they can count forward from the last counting word rather than re-counting the entire set. This is the strategy called **counting on,** and it is an important developmental milestone that comes after a prolonged period of counting all.

In kindergarten, many teachers begin working on the concept of counting by twos. The School Bus Song activity can be used for that concept as well. Simply add two photos to the bus for each verse. Remember that some children in the class may still need to count all of the photos to determine how many are represented. Both concepts (counting by twos and counting all) can be supported by counting each photo, but emphasizing the even numbers, in the following manner: one, *two*, three, *four*, five, *six*, seven, *eight*—two, four, six, eight. By pointing to the appropriate photograph for each number counted, the teacher connects the counting string to the appropriate quantities represented.

Number Sense Throughout the Curriculum

Highly effective teachers take every opportunity to reinforce number-sense concepts as they occur throughout the curriculum. In addition, through careful planning of the environment, they ensure that more opportunities will arise. Shown next are some suggestions for focusing on number sense throughout the preschool and kindergarten curriculum.

Art Area

Donald Crews (1986) has written a delightful counting book called *10 Black Dots* in which he uses from 1 to 10 black dots in each of his graphic illustrations. After hearing the story, children can create their own black-dot pictures. Teachers can use a 1-inch circle punch to precut circles from black construction paper. Children choose how many circles to glue onto their paper and use markers, crayons, or colored pencils to finish their pictures. Teachers can assemble the pictures into a class book or mount them on the wall to create a number collage.

Science Area

Buoyancy is an interesting science concept for young children to explore, and it involves mathematics. Teachers might place a tub of water in the science area, along with two plastic boats or other containers of different sizes. Children can add marbles to the boats and determine how many marbles it takes to sink each one. They will observe the boats floating lower and lower in the water as they lose buoyancy and eventually sink.

Dramatic Play Area

A pretend grocery store, ice cream shop, restaurant, or shoe store each provide opportunities for children to quantify and represent number as part of their play. As part of their plans for the area, teachers might supply price tags or menus, and either play money or actual coins. For preschool, teachers may use only pennies. In kindergarten, pennies, nickels, and dimes may be used. Teachers should expect some kindergarten children to struggle with the idea that one nickel is worth five pennies.

Gross Motor Area

Path games in the gross motor area help children construct the idea of a number line. All that are needed are a large die and something to form a path. In the outside area, a chalk path can be drawn on the blacktop or sidewalk. Inside, carpet squares or shapes cut from heavy paper can form the path. Children roll a large die (a fuzzy dashboard die works well) to determine how many spaces they are to hop along the path. When children use their own bodies as movers, they seldom rehop the space that they landed on to end their previous turn. They are much more aware of which spaces have already been counted.

Book Area

In addition to counting books, many excellent children's books reinforce number-sense concepts. The following are just a few of the growing list of books in this category:

The Doorbell Rang, Pat Hutchins (1986); *Benny's Pennies,* Pat Brisson (1993); *The Very Hungry Caterpillar,* Eric Carle (1994); and *How Many Seeds in a Pumpkin,* Margaret McNamara (2007)

UNIVERSAL DESIGN FOR SUPPORTING NUMBER SENSE

The activities presented in this chapter are self-leveling. This means that children at various stages of development can use the same materials in different ways. For example, children who are at the global, one-to-one correspondence, or counting stages of quantification all can play the math manipulative and grid games. In addition, teachers can vary the difficulty levels of the path games and card games by increasing or decreasing the sizes of the sets on the dice or cards. Because these games are so adaptable, they allow children at different cognitive levels to participate and demonstrate their knowledge. In other words, they offer multiple means of action and expression, an important UDL concept.

Teachers must always be aware of the particular needs of individual students. Adaptations to activities and materials should be made before they are presented to the class. For children who have visual difficulties, materials should be altered so that the mathematical properties can be felt as well as seen. For example, the spaces on path games can be outlined with puffy paint after the game board has been laminated, and dice with raised dots can be used. Puffy stickers mounted on 1-inch cubes make attractive dice that allow children with visual difficulties to participate along with their peers. Sometimes, children have coordination problems or extraneous movements that cause them to unintentionally knock game boards off the table. Simply taping the game board to the table will eliminate this problem. Children with autism or severe communication disorders may need help with turn taking. Cards with photographs of the children and their names can be mounted on a notebook ring. (First get permission from parents to photograph students.) Before the game begins, the teacher can remove the cards that are not needed. The children can flip the cards to indicate whose turn is next.

Affect has a strong impact on learning, and mathematics is known to create anxiety in many individuals. Teachers should be aware of this and watch for children who seem to avoid math or are uncomfortable during math activities. Sometimes, moving an activity to another area of the curriculum, such as creating path games in the gross motor area, is helpful. Embedding math throughout the curriculum provides children with various ways of acquiring information. This is another tenet of UDL.

Sometimes, children may lack the background to successfully engage in a particular math activity. This is more likely to be the case in kindergarten than in preschool, particularly if workbooks are used. The following example illustrates the problems that children in a structured kindergarten math environment may encounter.

EXAMPLE 2.6

Rachel, a student teacher in a kindergarten placement, worked patiently with Steve on his workbook assignment. At the top of the page was a list of the first 20 numerals. The assignment was to write the number that came before and after the numeral in each question, as in ____ 12 ____. Steve had no idea what numbers to write. Again and again Rachel showed him how to find the number on the number line and look to the left and right of the number to determine what to write on his paper. Nevertheless, each time Steve came to a new question, he was as lost as he was before. The rest of the class had all gone outside for recess while Rachel and Steve struggled on. Finally, Steve looked at Rachel and said, "My stomach hurts." At a loss as to what to do next, Rachel turned to her supervisor, who had been observing this interaction, and said, "I just don't know what to do." Her supervisor suggested that Steve's stomach might feel better if he could go out and play.

Once Steve had joined his friends on the playground, Rachel and her supervisor talked about Steve's math problems. With no concrete understanding of what the number line represented, Steve could not do the workbook assignment. Simply repeating what he could not understand would not help. The supervisor suggested that, with the cooperating teacher's permission, Rachel take some time each day to play math games with Steve. They would take turns rolling a die and taking counters so that Steve could become familiar with the relative magnitude of the numbers. Rachel's cooperating teacher was pleased with this idea, and the plan was implemented.

Three weeks later, Rachel's supervisor returned for another observation. Rachel happily reported that she and Steve had enjoyed playing math games in the hall. Steve's understanding of numbers had dramatically improved, and he was now able to complete his workbook assignments with little assistance.

Many kindergarten teachers feel pressed for time. They have much more material to cover than previous generations of teachers did, and curricula are often structured and required. This presents a dilemma, because not all children can gain access to the adopted curriculum. As in Steve's case, a small amount of time set aside each day for math games can dramatically increase a child's understanding. When math is integrated into other curriculum areas, the benefits begin to multiply.

FOCUSING ON MATHEMATICAL PROCESSES

The five mathematics process standards—Problem Solving, Reasoning and Proof, Communication, Connections, and Representation—are embedded throughout this chapter. As an example, the earlier Bathtub Game employs all five processing standards.

- *Problem Solving*—Each time children roll the die, they must solve a new mathematical problem that involves quantifying the number of dots on the die and creating an equivalent set of people. Children will select a strategy based on their current level of understanding.

- *Reasoning and Proof*—As children play the game, they may question one another about accuracy. This requires the other child to justify her response. Also, children may decide on a strategy to prove who has more people at the end of the game.

- *Communication*—The game offers children the opportunity to communicate mathematically. They may tell each other how many units they rolled on the die, discuss who has more or fewer people, or point out counting errors.

- *Connection*—The game connects directly to a favorite children's book, so children can see a connection between literacy and mathematics.

- *Representation*—In this game, children use objects (toy people) to represent mathematical problems.

ASSESSING NUMBER SENSE LEARNING

Ongoing, **formative assessment** is a critical part of mathematics instruction. By observing or joining in a game with young children, teachers can quickly determine children's current level of quantification and their understanding of counting principles. Rather than correcting children's errors, which are really just a reflection of their current developmental level, teachers can model the next step in development, talk about what they are doing, and encourage peers to discuss their problem-solving strategies. It should be remembered that development takes time and experience. Correcting young children's errors may have the adverse effect of discouraging them from participating in math activities or tackling only easier problems. Educators who want children to continue to enjoy mathematics and to take cognitive risks must allow them to make errors in the process.

The examples that follow are typical of the behaviors which children may exhibit when engaging in classroom math activities. From these examples, teachers can assess the children's current thinking and decide on appropriate teaching techniques to move them forward in their development.

EXAMPLE 2.7

Michael selects a snowflake grid game from the shelf. The game board has 12 spaces, each with an image of a snowflake. There is a basket of plastic snowflakes to use as cover-up pieces and a 1–3 dot die. Michael begins the game by placing one snowflake counter in each box in the first row of his card. Then he notices the die. He rolls it and points to the dots one at a time. Each time he points to a dot he takes a snowflake counter and places it on a snowflake picture.

This is an example of a child at the one-to-one level of quantification. The idea of applying a one-to-one relationship in a game situation may be new to him; at first he does not use the die. The teacher can begin by acknowledging this relationship. She might say, "I noticed that you took one snowflake for each dot on your die. I'd like to try that, too. Can I take a game board and play with you?" For the first few days, the teacher may be wise to continue to support one-to-one correspondence. Then she may decide to introduce counting *when it is her turn*. She may simply say, "1, 2, 3—I get to take 1, 2, 3 snowflakes." Although Michael likely will make note of this change, the teacher should expect him to continue using his own strategy until he has solidified the concept of applying counting in these situations.

EXAMPLE 2.8

Josh plays a grid and counter game that has 16 spaces and a 1–6 dot die. When he rolls 1, 2, or 3, he correctly counts the dots and takes the same number of counters. When he rolls 4, 5, or 6, he turns the die back to 1, 2, or 3 before taking his counters.

This is an interesting situation. Usually, when children fix the die, they do it to get a higher number rather than a lower one. Josh demonstrates the counting level of quantification for numbers 1–3. For these small numbers, he has demonstrated stable order counting (correct number names and order), one-to-one correspondence (one number word for each object), and cardinality (the last number counted is how many counters he takes). Why, then, does he turn the die back to smaller numbers? It could be that he previously had made a mistake when counting larger numbers and now wants to play it safe. His behavior also may indicate that he does not know the counting words for numbers above 3. In either case, the teacher should allow Josh to continue doing what he is comfortable with, and model counting the higher numbers on her own turn. Another strategy would be to encourage another child who counts higher quantities to play with Josh. This child may question Josh's fixing of the die and force him to work with higher numbers.

EXAMPLE 2.9

Sanjay and Maria are playing a path game. When Sanjay rolls a 5 on the die, he points to each dot and counts 1, 2, 3, 5, 6. He then moves 5 spaces along the path as he counts 1, 2, 3, 5, 6. Maria tells Sanjay he got only a 5, so he needs to move back 1 space. Sanjay insists he moved the right number of spaces.

This is an interesting dilemma for these young mathematicians; both are partially correct and partially incorrect. This is where the Reasoning and Proof process standard can be helpful. The teacher might start by asking Maria to explain what she means. She will likely point out that Sanjay left out the number 4 when he was counting. If the teacher then asks Maria to demonstrate how Sanjay should have moved, Maria will discover that she ends up at the same place at which Sanjay sat his mover. By using her fingers, the teacher can demonstrate that each child said five number words. Interactions such as this help both children. Sanjay can change his number word string, and Maria will continue to ponder why you can leave out the number 4 and still end up on the right space.

These assessment examples illustrate that no child's learning is stagnant. Each child should continue solidifying knowledge and moving forward.

SUMMARY

This chapter has covered a great deal of information related to number-sense concepts and children's development. Future chapters will build upon this information, so it is important that it become a regular part of teachers' thinking as they plan, implement, and evaluate mathematics instruction.

Two key components of the number-sense standard for preschool and kindergarten children are quantification strategies and understanding of counting. At first it may be difficult to separate the two. Quantification refers to how children decide on the magnitude of a set of objects, and there are three levels of understanding. At first, young children use visual perception to make a global decision, such as "a lot" or "not so much." At the next level, children understand that they can match the original set of objects by using one new object, such as their finger or a counter, to represent each object in the original set. Finally, at the third level, children understand that they can use counting to quantify because the last object they count represents the entire group.

Even after children begin to use counting as a quantification strategy, developmentally related errors are typical. Children must construct three "how to count" principles in order to be accurate when counting. They must use the correct counting words in the correct order each time they count; they must count each object once, without skipping over or

SUMMARY (*continued*)

re-counting any; and they must understand cardinality, or the fact that the last number counted represents the total.

Teachers should regularly plan math activities for individual children, small groups, and the entire class. These activities should be incorporated into all areas of the curriculum so that children have multiple ways to think about and represent mathematical problems. Assessment should be ongoing, with teachers regularly observing children's use and understanding of number-sense concepts.

ON YOUR OWN

■ Select an activity from this chapter and discuss how it encourages children to use the five processing standards.

■ Use natural materials or materials commonly found around the home to design a number-sense activity.

■ Observe a group of children engaging in play. How do number-sense concepts emerge as part of the play?

REFERENCES

Aunola, K., Leskinen, E., Lerkkanen, M.K., & Nurmi, J.E. (2004). Developmental dynamics of math performance from pre-school to grade 2. *Journal of Educational Psychology, 96,* 699–713.

Baroody, A.J., Lai, M., & Mix, K.S. (2006). The development of young children's number and operations sense and its implications for early childhood education. In B. Spodek & O.N. Saracho (Eds.), *Handbook of research on the education of young children* (pp. 187–221). Mahwah, NJ: Erlbaum.

Brisson, P. (1993). *Benny's pennies.* New York: Doubleday Book for Young Readers.

Carle, E. (1994). *The very hungry caterpillar* (25th anniversary edition). New York: Philomel.

Clements, D.H., & Sarama, J. (2007). Early mathematics learning. In F.K. Lester, Jr. (Ed.), *Second handbook of research on mathematics teaching and learning* (p. 462). Reston, VA: National Council of Teachers of Mathematics.

Clements, D.H., Sarama, J., & DiBiase, A.M. (2004). *Engaging young children in mathematics: Standards for early childhood mathematics education.* Mahwah, NJ: Erlbaum.

Crews, D. (1986). *Ten black dots.* New York: Greenwillow.

Gelman, R., & Gallistel, C.R. (1978). *The child's understanding of number.* Cambridge, MA: Harvard University Press.

Griffin, S., Case, R., & Siegler, R.S. (1994). Rightstart: Providing the central conceptual prerequisites for first formal learning of arithmetic to students at risk for failure. In K. McGilly (Ed.), *Classroom lessons: Integrating cognitive theory and classroom practice* (pp. 25–49). Cambridge, MA: MIT Press.

Hutchins, P. (1986). *The doorbell rang.* New York: Greenwillow.

Kamii, C. (1982). *Number in preschool and kindergarten: Educational implications of Piaget's theory.* Washington, DC: National Association for the Education of Young Children.

Kamii, C. (1994). *Young children reinvent arithmetic: 3rd grade.* New York: Teachers College Press.

Kamii, C. (2000). *Young children reinvent arithmetic: Implications of Piaget's theory* (2nd ed.). New York: Teachers College Press.

Kamii, C. (2004). *Young children reinvent arithmetic: 2nd grade* (2nd ed.). New York: Teachers College Press.

Kato, Y., Kamii, C., Ozaki, K., & Nagahiro, M. (2002). Young children's representations of groups of objects: The relationship between abstraction and representation. *Journal for Research in Mathematics Education, 33*(1), 30–46.

Krasa, N., & Shunkwiler, S. (2009). *Number sense and number nonsense: Understanding the challenges of learning math.* Baltimore: Paul H. Brookes Publishing Co.

McNamara, M. (2007). *How many seeds in a pumpkin?* New York: Schwartz & Wade.

Moomaw, S. (2008). *Measuring number sense in young children.* (Doctoral dissertation, University of Cincinnati). Retrieved from http://www.ohiolink.edu/etd/.

Moomaw, S., & Hieronymus, B. (1995). *More than counting.* St. Paul, MN: Redleaf Press.

Moomaw, S., & Hieronymus, B. (1999). *Much more than counting.* St. Paul, MN: Redleaf Press.

National Council of Teachers of Mathematics. (2000). *Curriculum and evaluation standards for school mathematics.* Reston, VA: Author.

National Research Council, Committee on Early Childhood Mathematics. (2009). (C.T. Cross, T.A. Woods, & H. Schweingruber, Eds.). *Mathematics learning in early childhood: Paths toward excellence and equity.* Washington, DC: The National Academies Press.

NCES (2009). *Highlights from Trends in International Mathematics and Science Studies (TIMSS) 2007.* United States Department of Education: NCES 2009-001 Revised.

Perry, B., & Dockett, S. (2002). Young children's access to powerful mathematical ideas. In D.L. English (Ed.), *Handbook of international research in mathematics education* (pp. 81–111). Mahwah, NJ: Erlbaum.

Piaget, J. (1952). *The child's conception of number.* New York: Norton.

Ramani, G.B., & Siegler, R.S. (2008). Promoting broad and stable improvements in low-income children's numerical knowledge through playing number board games. *Child Development, 79*(2), 375–394.

Thomson, S., Rowe, K., Underwood, C., & Peck, R. (2005). *Numeracy in the early years.* Melbourne, Australia: Australian Council for Educational Research.

U.S. Department of Education, N.C.E.S. (2000). *The condition of education 2000* (p. 7). Washington, DC: U.S. Government Printing Office.

Vygotsy, L.S. (1978). *Mind in society. The development of higher psychological processes.* Cambridge, MA: Harvard University Press.

Wadsworth, B.J. (1989). *Piaget's theory of cognitive and affective development* (4th ed.). New York: Longman.

Whyte, J.C., & Bull, R. (2008). Number games, magnitude representation, and basic number skills in preschoolers. *Developmental Psychology, 44*(2), 588–596.

Wood, A. (2005). *King Bidgood's in the bathtub.* Orlando, FL: Harcourt.

Young-Loveridge, J.M. (2004). Effects on early numeracy of a program using number books and games. *Early Childhood Research Quarterly, 19,* 82–98.

3

Developing Concepts About Arithmetic Operations

1 and 1 are 2, 2 and 2 are 4 . . .

That's 2 for you, 2 for you, and 2 for me—6 cookies.

We'll divide them. Here's some for you and some for me.

Addition, subtraction, multiplication, and division are arithmetic operations that are a focus in Grades 1–4. Important foundational concepts related to all four operations, however, begin long before these grades, in the play experiences of young children. Friends may want to find out how many stickers they have when they put two more into their collection, so they count all of their stickers (addition). Dad may require that siblings share a toy car with baby, so they have to re-count how many cars they have left (subtraction). The teacher may ask students to bring enough mittens from the mitten bin for three children, so they think about repeated addition (multiplication). A group of four friends may decide to divide up the toy fish in the water table, so they put a pile of approximately the same size into each child's pail (division). All of these interactions provide opportunities for children to think about and experiment with the four arithmetic operations. By the time these are presented formally in the primary or later elementary grades, these children have solid mathematical concepts to build upon. Supporting the development of arithmetic operations is another important component of the Number and Operations standard (NCTM, 2000).

THE OPERATIONS COMPONENT OF THE NUMBER AND OPERATIONS STANDARD

Beginning concepts of arithmetic operations, including how they relate to one another, are included in the NCTM Number and Operations standard for children in preschool through second grade. Teachers of preschool and kindergarten children initially may be surprised that addition, subtraction, multiplication, and division are concepts that begin to develop during these early years. Close observation, however, reveals that children begin solving problems related to all four arithmetic operations during preschool and kindergarten. Of course, their strategies are not as sophisticated as those of older children; nevertheless, like all mathematicians, young children apply the foundational knowledge that they have already developed to more challenging and interesting problems. The examples that follow show children working to solve addition, subtraction, multiplication, and division problems that their teachers initially thought might be beyond their reach.

EXAMPLE 3.1

Silje (age 5) rounded up some friends to play a target game in the gross motor room. The target had three concentric circles, each a different color, and beanbags to throw at it. The teacher had hoped that the children might assign points, such as 1, 2, and 3, to the circles on the target and keep score. Silje did indeed assign points to the circles, but not what her teacher had anticipated. Instead, she and her friends decided that the biggest circle would be worth 10 points, the middle circle 20 points, and the smallest circle 30 points. The teacher stifled the urge to suggest smaller point values, and the girls proceeded with their game.

Silje's friend Anna went first. On her first throw, Anna hit the small circle and announced that she had 30 points. Then she hit the large circle for 10 points and the middle circle for 20 points. To determine the score, Silje started with 30 and counted 10 more on her fingers to get 40. After some thought, she continued counting 20 more. Silje counted accurately to 59 but was not sure what the final number should be. The teacher supplied the final number in the count string—60.

This example demonstrates the relationship between counting and early addition. Even though the numbers were large, Silje understood that she could use her substantial counting skills to add the point values together. She showed strong development of the stable order, one-to-one correspondence, and cardinality counting principles, which allowed for accuracy in her attempt to solve this problem. In addition, Silje was able to count forward from one point value to the next as she combined the numbers.

The problem that Silje created for herself would be too difficult for many of her classmates; hence, the teacher's initial hesitation. Children are eager to tackle difficult problems, however, when they can use their own strategies and are not being evaluated. The teacher was wise to allow the girls to tackle this problem. Had it proven to be too difficult to solve, the teacher could have suggested smaller point values as an intermediary step to thinking about the larger numbers.

EXAMPLE 3.2

Jesse (age 4) had 6 sight words in his word bank. The words were written on note cards and held together by a notebook ring. Each day, Jesse had the opportunity to add a new word to his word bank. On this particular day, he asked for the word dinosaur, *which his teacher printed on a note card. "How many words do you have now?" asked the teacher. Jesse counted his cards and replied that he had 7 words.*

Jesse was eager to get 10 words in his word bank because then his teacher would make a game for him that used the words. Now he wondered how many more words he needed to reach 10. "How could you figure that out?" asked his teacher. Jesse spread out his cards and re-counted them, again ending up with 7. This time, though, he continued counting,

> EXAMPLE 3.2 *(continued)*
>
> *placing one finger next to his row of cards each time he said another number word. Then Jesse looked at the three fingers he had extended and proudly told his teacher that he needed only 3 more words to get to 10.*

This scenario illustrates the relationship between addition and subtraction. Although the problem could be solved by subtracting 7 from 10, it also can be solved by the method that Jesse devised of adding on to the 7 until he reached 10. This is the type of problem in which the start value (7) and the end value (10) are known, but the change is unknown (Carpenter, Fennema, Franke, Levi, & Empson, 1999). Like Silje in the previous example, Jesse used counting to solve the problem; however, unlike Silje, Jesse needed to re-count the first quantity before continuing his counting string.

> EXAMPLE 3.3
>
> *Two children were helping their preschool teacher make white frosting for the cupcakes that their class was baking. "Can we make the frosting a different color?" asked Shawn. "What color do you want?" the teacher replied. The two children could not decide which color the class would like best, so they decided to make yellow, blue, and pink frosting. The teacher reminded the children that the rule when cooking was that each child had to have his or her own spoon for each bowl. "Shawn," she said, "you can go to the kitchen and ask the cook for spoons. How many will you need?" Shawn looked intently at the three bowls of frosting. Then he tapped two times in front of each bowl as he quietly counted—1-2, 3-4, 5-6. "We need to ask for 6 spoons," he told the teacher.*

One way of conceptualizing multiplication is as repeated addition. When each person in a group (in this case, 2) needs the same number of objects (as in 3 spoons), the problem can be solved by counting each object the same number of times as the number of people. Shawn realized that he needed one spoon for himself and one for his friend for each of the three bowls of frosting. By counting twice for each bowl, he solved a multiplication problem.

> EXAMPLE 3.4
>
> *Brad was reading the book* The Doorbell Rang, *by Pat Hutchins (1986), to a group of children in his kindergarten class. At the beginning of the story, 2 children had 12 cookies to share. Brad used 2 toy figures to represent the children and 12 flat marble chips to represent the cookies. "How should I divide up the cookies so it's even?" Brad asked. The children agreed that he should give one chip to each toy person until they were all used up. "Does each person have the same amount?" Brad asked, after dealing out the chips. "Count them," the children recommended. Together they counted 6 chips next to each person, and the children agreed that each person had the same number of cookies.*
>
> *Brad continued the story, in which two friends had now come to the door. This meant that the cookies needed to be shared among 4 people. Brad placed 2 more toy figures on the carpet in front of him. "Now what should we do?" he asked. Several children thought that Brad should take 1 cookie from each of the original 2 people and move them over to the 2 new people. Brad followed through with this idea. "Do they each have the*

(continued)

same number of cookies?" he asked. "No, those 2 people have more," several children said while pointing to the original 2 people. "Well, what should I do next?" Brad asked. "Take 1 more away from those first 2 people and give it to the new people," Sherry advised. Brad did so, leaving the 2 original people with 4 cookies and the 2 new people with 2 cookies. "Okay," Brad said. "Is it even now?" At this point there was disagreement. Some children were satisfied now that each person clearly had more than 1 cookie, but other children demanded that the cookies be re-counted. Together, Brad and the children counted each person's cookies. "See," said Phil. "Those 2 people have 4 cookies and those other 2 people only have 2 cookies."

"It looks like we still have a problem," Brad said. "What do you think we should do?" There was silence while everyone contemplated the scenario. Then Kenan, Phil, and Sherry each had the same idea. "Move 1 from that pile to that pile," said Sherry, pointing first to a pile of 4 cookies and then to a pile of 2 cookies. Brad followed her directions. "Then take 1 from that pile and move it to this other pile," said Kenan, pointing first to a pile of 4 cookies and then to the same pile in which Brad had just placed an extra cookie. This left the 2 original people with 3 cookies each, 1 of the new people with 4 cookies, and the other new person with 2 cookies. "Oh no," said Sherry. "That didn't work. This person has too many." She pointed to the pile of 4 cookies.

The children looked very confused. "Should we re-count all the cookies?" Brad asked. The children agreed that this was a good idea. After they had counted successive piles of 3, 3, 4, and 2 cookies, Phil knew what to do. "Take 1 cookie from the 4 and put it with the 2," he said. Once this was accomplished, the children agreed that the cookies were at last divided equally.

In this example, children used repeated subtraction to move objects from larger groups into groups with fewer objects. Because Brad modeled exactly what the children described, they could determine when they had made a mistake and think about how to correct it. Play in the block or dramatic play areas can provide similar opportunities for children to use arithmetic operations if the teacher provides the necessary props and poses mathematical questions. In both areas, sharing of materials is often an issue. If children are allowed to solve these problems together, rather than responding to rules set forth by the teacher, they can develop foundational concepts of division. Children may advance from an initial, global response, in which everyone gets some of the items that they are attempting to share, to a one-to-one correspondence relationship, in which objects are dealt out like cards to each individual. With teacher support, children may even quantify the disputed set of objects and decompose it into equal parts for each individual.

THE DEVELOPMENT OF ARITHMETIC OPERATIONS—ADDITION AND SUBTRACTION

The previous examples show that young children can solve problems involving all four arithmetic operations and should be given the opportunity to do so, as this increases their mathematical reasoning skills. However, the first two arithmetic operations that children are formally introduced to in elementary school are addition and subtraction. For this reason, the Number and Operation standard focuses primarily on those two operations for young children, and this chapter does the same.

Addition

As the previous examples illustrate, young children extend their use of counting to gradually incorporate arithmetic operations. Research has shown a hierarchy in children's responses to addition problems (Clements & Sarama, 2007). Children initially rely on counting strategies that require the use of countable objects such as fingers. They start by counting individual sets for the addends and then count them all together. For example, they might add sets of 3 and 4 objects by counting 1, 2, 3, (pause) 1, 2, 3, 4, (pause) 1, 2, 3, 4, 5, 6, 7. This strategy is often referred to as counting all (Baroody & Tiilikainen, 2003; Clements & Sarama, 2007). However, preschool teachers and researchers who have implemented a mathematics game curriculum, in which children quantify dot dice before taking counters or moving along a path, have noted that many quickly develop a true counting all strategy in which they count both sets together to get the sum without having to first count them separately (Moomaw & Hieronymus, 1995, 1999). For example, children who roll dice with 2 and 5 dots, respectively, would simply count 1, 2, 3, 4, 5, 6, 7. This is sometimes referred to as a **shortcut-sum** strategy (Baroody & Tiilikainen, 2003; Clements & Sarama, 2007). It is a notable development because it shows that children realize that they can get the sum by simply counting all the dots together, without having to quantify the individual sets. Eventually, children begin to count on from one of the addends. This means that they recognize (subitize) one of the sets and count forward from that number, without having to recount it. For example, a child adding sets of 3 and 4 might immediately say 3 for the first set and then continue 4, 5, 6, 7. Children at this stage are able to view the component addends as part of a larger sum (Kamii, 2000).

Teachers are more likely to see counting-on strategies emerge in kindergarten than in preschool, although sometimes an experienced or older preschool child may demonstrate this level of thinking. Young children seem to use a shortcut-sum strategy for a long period when combining sets, even after they can subitize at least one of the addends. Teachers may feel frustrated when a child who correctly identifies the quantities in two sets insists on counting all of the objects, starting with 1, to get the total, instead of counting on from the second set. Kamii clarifies that addition is the mental act of combining two wholes (the addends) to get a higher-order whole (the sum). Part–whole relationships are difficult for young children to comprehend. They have trouble thinking about an addend, which has its own cardinal value, as also being part of a new set with its own cardinal value. To sort out these relationships, children must count all of the objects. When they no longer need to count all, they will count on.

After many experiences with combining two sets, children begin to remember addition combinations (Kamii, 2000). Children who have repeated opportunities to compose sets, as they do with a game-based curriculum, do not need to explicitly memorize addition combinations; they simply remember them, as a part of their conceptual development. The combinations that are most easily remembered by young children appear to be the doubles, particularly 1+1, 2+2, 3+3, and 5+5. This has been noted by researchers (Kamii, 2000) as well as classroom teachers. Next, children learn +1 combinations, followed by +2. For this reason, 6+1 may be an easier combination to remember than 2+3, even though 7 is a higher sum than 5.

Subtraction

Subtraction is a more difficult concept for young children than addition. In essence, they must separate a whole number into two parts and quantify the part that is not given. For example, to subtract 4 from 7, they must quantify 7, remove 4 items, and then count what is left. Young children may approach a subtraction problem in one of two ways: 1) by modeling the problem directly or 2) by adding on from one of the parts to get the total. In either case, counting is involved. Teachers may observe children modeling subtraction when they sing a song such as "Five Little Ducks." If toy ducks are available, children may line up 5 ducks, remove 1 duck as they sing each verse, and then count the remaining ducks. If objects are not available, they may use their fingers to represent the ducks. This type of problem is an example of subtraction when the result is unknown. Rather than counting backward to find the result, preschool and kindergarten children usually count the remaining objects to find out what is left.

A second type of subtraction problem involves a change that is unknown. Suppose that Andre has 5 cars. He leaves briefly, and when he returns, only 3 cars are left. The teacher asks him how many cars are missing. Knowing that he had 5 cars to begin with, Andre may count the 3 remaining cars (or simply say 3) and extend 1 finger for number 4 and a second finger for number 5 to determine that 2 cars are missing. He uses forward counting from the set that he can see to determine the size of the set that he cannot see. These examples illustrate that, although subtraction may be more difficult than addition, young children can invent strategies to solve subtraction problems. By arranging and rearranging a small group of objects, such as ducks or cars, children are actually composing and decomposing numbers. These play opportunities help them construct the relationship between addition and subtraction.

Representation

Many children learn to recognize and name some numerals in preschool, and this ability expands during kindergarten. In one study, 93% of kindergarten children could read and write the numerals 3 and 9 by the end of the school year (Clements & Sarama, 2007). Yet, being able to name and produce numerals is very different from understanding what they mean. Many young children who can name numerals cannot represent their cardinal values with objects; in other words, they cannot hold up three fingers or show three blocks to represent the numeral 3. Even children who can do this tend not to use numerals to represent quantities. In a study by Kato, Kamii, Ozaki, and Nagahiro (2002), children between the ages of 3 years 4 months and 7 years 5 months were asked to represent objects placed on a table, such as 3 spoons or 8 blocks. Over half of the children who could write numerals did not use them to represent quantities of objects; instead, they represented the objects in a one-to-one correspondence fashion by either drawing every object or writing a numeral for each object rather than the total.

Children first use numerals to label groups of objects or to represent a quantity that they have counted. Kamii (2000) reports children making the transition into this use of numerals during first grade. However, this does not mean that children can translate this use of numerals into representation of addition. In a study conducted by the author of this book, 40 preschool children (ages 3–5) who could use numerals to represent quantities played a game in which they drew two cards, each containing either a set of dot stickers or a numeral, and took dinosaur counters to represent the total. The three additive conditions investigated in the study were, 1) adding two symbols (dots on both cards), 2) adding two signs (numerals on both cards), and 3) adding one symbol and one sign (one card with dots and one with numerals). Small addends (2, 3, and 4) were used in all conditions. The results reflected a significant decrease in children's scores when a numeral appeared on either of the cards. The study therefore showed that whereas young children who have had repeated experiences with quantification are ready to move into addition, they may be held back conceptually if quantities are represented by numerals.

ADDING OPERATIONS TO THE NUMBER-SENSE CURRICULUM

Addition and subtraction are natural outgrowths of counting. Children experience many situations each day that involve these two operations; however, adults often do not take advantage of these opportunities to encourage children to think about arithmetic operations. In addition to daily living and play experiences that involve addition and subtraction, games that involve rolling two dice or drawing two cards with concrete sets (e.g., dots or stickers rather than numerals) allow children to progress naturally into addition, first by counting all, then by applying a short-cut sum strategy, and finally by counting on and even remembering addition combinations. Math games that use concrete objects can also support beginning concepts of subtraction. Teachers can use group time to introduce modeling of addition and subtraction, such as by dramatizing songs or stories that involve these operations. For these reasons, curriculum planning in this chapter is again organized around the following areas: math talk, individual and small-group activities, large-group experiences, and operations integrated throughout the curriculum.

Math Talk

Snack and lunch time provide numerous opportunities for adults to integrate addition and subtraction into their interactions and conversations with children. Eating accounts for a significant portion of the day in many preschool and kindergarten programs, and this learning time should not be lost. In particular, children who are labeled "at risk" due to poverty, second-language acquisition, or disabilities need maximum opportunities to talk about math in meaningful situations. The examples that follow illustrate how arithmetic operations can form a natural and recurring part of meal conversations.

EXAMPLE 3.5

At the lunch table, Alex asks for more grapes. His friend Lucy says that she wants more, too. Rather than require Lucy to first eat the 2 grapes she already has, the teacher asks, "How many do you have already?" After Lucy responds "2," the teacher says, "Okay. I'll give you 2 more. Then you can see how many 2 grapes and 2 more grapes is." Lucy happily counts 4 grapes on her plate.

Knowing that 2 + 2 is one of the earliest addition combinations for children to learn, the teacher chooses that combination to insert into this naturally occurring situation. The language that the teacher uses encourages Lucy to think about 4 as being composed of 2 units and 2 more units.

EXAMPLE 3.6

Lamonte announces, "I have 5 carrot sticks. Just like me. I'm 5." The teacher responds, "That's cool. How many will you have after you eat 2?" Lamonte picks up 2 carrots and looks at the remaining carrots on his plate. "3," he quickly responds. "Wow," the teacher follows up, "so if you have 5 carrots and take 2 away, you have 3 left."

In this case, the teacher introduces subtraction because it represents what is about to happen. Lamonte has quantified a set of carrots, will soon eat some, and will have a smaller number remaining. By framing the situation in mathematical terms, the teacher encourages Lamonte to think about the decomposition of 5. He does this rather easily with the actual carrots in front of him; however, Lamonte would likely not have thought about the situation in mathematical terms if the teacher had not introduced it into the conversation.

EXAMPLE 3.7

"We're having blueberry muffins for snack," announces the teacher, holding up a basket with mini-muffins.

"Can we have 2?" asks Megan.

"I don't know," replies the teacher. "How many muffins have to be in the basket for each of you to have 2?"

Megan looks at the 4 children at the table. Then, pointing her finger twice at each child, she counts 1,2—3,4—5,6—7,8. We need 8."

The teacher lines up the muffins on the table. Together, he and the children count 10 muffins. "Is that enough?" he asks.

"Yes," Megan answers. "We already said 8 muffins when we counted to 10."

Not all of the children seem so sure. The teacher places 2 muffins on each child's plate, counting as he does so. "Did we have enough?" he asks.

"Yes!" the children reply happily.

"How many muffins are left over?" the teacher asks.

The children respond that there are 2 left. "You can have those, Mr. Jaumall," Precious says.

"Thanks, Precious," the teacher replies, "but I'm going to eat later. How can we divide these up among the four of you?"

The children look stumped. "There's not enough," says Megan.

"Break them up," says Amber. "That's what my mom does."

"Good idea," the teacher says. "I'm going to cut each muffin in half. Now, let's count the pieces." Together they count 4 pieces.

"That's enough," Megan exclaims.

"You each get 2½ muffins," Jaumall summarizes.

Sharing snack items is another familiar experience for young children. In this case, the teacher capitalized on the opportunity to introduce mathematical problem solving that incorporated counting; repeated addition, or multiplication; subtraction; and even division involving fractions. Megan incorporated repeated addition (i.e., multiplication) when she determined that 2 muffins for each of 4 children would be 8 muffins. The teacher introduced counting and comparison of sets when he helped the children count the muffins and asked whether 10 muffins is more than 8. He verified Megan's answer when he carefully counted the muffins as he distributed them. Then the teacher briefly focused on subtraction when he drew attention to the remaining muffins. Finally, he introduced division involving fractions for the sharing of the remaining 2 muffins. Fractions are not included in the mathematics core standards until third grade; however, mathematics educators recommend that they be informally introduced at much younger ages so that children have a foundation for later instruction (Fennell, 2010).

Opportunities to engage in discussions that involve arithmetic operations occur across the curriculum. Examples will be highlighted later in this section. In addition, math talk is a critical component of small- and large-group interactions and is included with each activity in those sections.

Individual or Small-Group Activities

Many of the activities introduced in Chapter 2 can easily be adapted to incorporate addition or subtraction. Simply adding a second die to a game introduces the concept of combining sets, or addition. Examples of manipulative and path games that encourage addition or subtraction are included in this section, along with activities that focus children's attention on composing and decomposing sets. (For many more examples of similar games and activities, see Moomaw & Hieronymus, 1995, 1999.)

Math Manipulative Activities

Because math manipulative games incorporate objects to represent numerical sets, they encourage children to focus on the composition and decomposition of numbers, an important component of addition and subtraction. For example, a child who is playing with 5 toy eggs in a nest can organize them into groupings of 1 and 4, 2 and 3, 3, and 2, or 4 and 1. Later, in the early elementary grades, these combinations may be organized into "fact families" for addition and subtraction and represented as shown here:

$4 + 1 = 5$	$5 - 4 = 1$
$1 + 4 = 5$	$5 - 1 = 4$
$3 + 2 = 5$	$5 - 3 = 2$
$2 + 3 = 5$	$5 - 2 = 3$
$5 + 0 = 5$	$5 - 0 = 5$
$0 + 5 = 5$	

Math manipulative games and activities also allow children to directly model addition and subtraction problems. To draw on the bathtub game example from Chapter 2, if 3 toy people are in the tub and a child puts 2 more in, the child can observe that there are now 5 people in the tub. Conversely, if 2 are removed, the total goes back to 3 people. Materials such as the bathtub game help children construct the mathematical relationship between addition and subtraction.

ACTIVITY 3.1

Eggs in the Basket

Materials

The following materials are needed for this activity:

- 12 small baskets
- Sponge insert, glued to the bottom of the basket, as shown
- 42 small plastic eggs
- 2 small bowls, to hold 21 eggs each for the start of the activity
- number cards attached to each basket, with numerals and dots from 1 to 6
- 1–6 dot die

Description

This activity, which focuses on the composition of numbers, can be played by an individual child or by two children as a game. Individual children can use the numbered baskets to determine how many eggs to place in each. Dots are included along with numerals so that children who do not yet recognize numerals can determine how many eggs to place in each basket. Once the eggs are in the baskets, children can move them from side to side to create different representations of the cardinal number. For example, the basket labeled "3" could have eggs arranged in four different ways: 1) 3 eggs in front of the divider and none behind it; 2) 2 eggs in front of the divider and 1 behind it; 3) 1 egg in front of the divider and 2 behind it; and 4) no eggs in front of the divider and 3 behind it. Older preschool or kindergarten children may want to draw or write the possible combinations for each number.

Two children can use the materials in a game format. Each child starts with a set of baskets (labeled 1–6) and a bowl of 21 eggs. Children take turns rolling the die to determine which basket to fill until all 12 baskets have the requisite number of eggs.

Math Discussions

This activity lends itself to discussions about number composition. Later, children will use this understanding to support addition. If a child is playing alone, the teacher might start a conversation by asking, "How did you decide how many eggs to put in this basket?" The conversation may continue with prompts from the teacher, such as the following: "This basket has 5 eggs, 2 on this side and 3 on this side. If you move 1 egg to the other side, will there still be 5 eggs?" A follow-up comment could be this: "I see that 2 eggs and 3 eggs is 5 eggs, and 1 egg and 4 eggs is 5 eggs also. That's pretty neat."

When two children play this activity as a game, the teacher could ask them to compare how their sets for various numbers look. For example, the teacher might say, "Look. Freddy's number 4 basket has 2 eggs on each side, but Monique's number 4 basket has 3 eggs on one side and 1 egg on the other side. See? Are there 4 eggs in each of your baskets?" Discussions such as these help children not only construct and deconstruct numbers, but also begin to think about the order-irrelevant counting principle (see Chapter 2). They gradually begin to realize that, no matter in which order they count the eggs in their baskets, the total remains the same.

ACTIVITY 3.2

Socks in the Laundry

Materials

The following materials are needed for this activity:

- Clothesline frame (see directions that follow)
- 20 small socks, cut from colored felt
- 20 small craft clothespins
- 1–3 spinner, made with sock stickers and a blank spinner
- 4 small baskets, to hold 10 socks and 10 clothespins for each player

Description

To make the clothesline frame, drill a $3/4$-inch-diameter hole into each end of a wooden base, approximately 15 × 4 inches. Cut two 8-inch lengths of $3/4$-inch-diameter dowel. Drill a $1/4$-inch-diameter hole through each dowel near the top. Insert a 5-inch length of $1/4$-inch-diameter dowel into each hole, center it, and glue it in place. Then glue the $3/4$-inch-diameter dowels into the holes in the base. When the glue has dried, stretch plastic-wrapped wire or pipe cleaners between the crossbars to form 2 clotheslines. Tape them in place if necessary.

Although this activity can be investigated by an individual child, it is best when played by two children sitting opposite one another. Children spin the spinner to determine how many socks to hang on their clothesline. With each turn, children can add a new set of socks to their clothesline. This gives them experience with combining sets and counting to determine the new total. The first child to have 10 socks wins round one of the game. That child must wait until the second child also has acquired 10 socks before continuing the game.

For part two of the game, children continue taking turns with the spinner, but this time they remove socks from the clothesline and place them back in the basket. This gives them practice with modeling subtraction. The game ends when all of the socks are back in the baskets.

Math Discussions

This game meets the needs of children at many different mathematical levels and supports many number-sense concepts. The content of math discussions, therefore, will depend on the thinking levels of the children playing the game. For many preschool children and some kindergartners, the focus may be on creating and comparing sets as they quantify the number of socks on the spinner and hang an equivalent number on the line. The design of the clothesline frame makes it easy for children to use one-to-one correspondence to compare how many socks they each have. Of course, some children may be fooled by their perceptions if one row of socks stretches farther than another row with the same number of socks. This is why math discussions are so important. Following are some examples of comments or questions that teachers might use to support learning at this level:

- *This looks like a fun game. Who has more socks so far? How can you tell?*
- *These socks are all lined up, aren't they? I can see that Max has 2 more socks than Victor. What does Victor need to get on the spinner to have more socks than Max?*
- *Victor seems to have run out of socks. How many are on your clothesline, Victor? Max, how many are left in your basket?*

For children who can easily quantify small amounts and make set comparisons, focus should shift to addition and subtraction. Here are some examples of teacher questions and comments that support such a focus:

- *You had 7 socks on your line. Now you have 1 more. How many is that?*
- *How many more socks do you need to get to 10?*
- *Carol says that she found out how many more she needs to get to 10 by looking in her basket. Carol, can you show us how you know that the socks on the line and the socks in your basket add up to 10?*
- *I got 2 and 2 and 3 on my turns. Help me figure out how many socks I have.*
- *You had 10 socks, right? Now you get to put 3 back in the basket. How many will be left?*
- *I have 4 socks left. Can I get them all back into the basket with one more spin?*

Because all of the topics discussed in these examples focus on the game, children generally are quite interested in discussing them. Whereas each comment or question draws attention to addition or subtraction, children can use a variety of strategies to determine the answer. For example, a child who wants to find out how many more socks he needs to get to 10 can count on from the number he has on the clothesline or count the remaining socks in the basket. Whichever strategy they choose, children have the socks available to model the problem.

Pairs of Socks— Counting by 2s

Materials

The following materials are needed for this game:

- ■ Small felt socks, from the previous activity
- ■ Shelf extender (as pictured)

Description

As a final extension of the sock game, children can group the socks in pairs. A small shelf extender can be substituted for the original frame if desired. Children can hang a pair of socks on each bar of the extender, which makes counting socks and pairs of socks easier.

Math Discussions

Young children often are quite perplexed to find that there are more socks than pairs of socks, so this is an obvious topic for conversation. Being able to see the socks, particularly if they are paired by color or design, helps children visualize and understand the relationship between pairs and individual units. The teacher can model counting the individual socks by emphasizing the even numbers, as in "1 **2**, 3 **4**, 5 **6**, 7 **8**, 9 **10**." Then, pointing to the pairs of socks, she can say, "That's 2, 4, 6, 8, 10."

Grid Games

The grid games explored in Chapter 2 can evolve into addition and subtraction games as children advance in their development. Simply adding a second die introduces the idea of combining sets, or addition. Of course, extra counters will be needed to accommodate the larger quantities rolled on two dice. In a game such as the Dog House Game discussed in Chapter 2, the rules can be modified so that 2 or 3 dogs go into each dog house, rather than 1 dog, as in the original game. If there are not enough dogs, cutouts of dog bones can be used instead. The grid game described next is specifically designed to focus on the composition and decomposition of sets.

Shop and Save

Materials

The following materials are needed for this game:

- ■ 2 grid boards (10 × 4 inches), each divided into 10 sections, or "boxes," with a sticker of a toy or food item in each box

- 2 small piggybanks, jars, or coin purses
- 40 pennies or toy coins in a bowl
- 40 green index cards, each with 3, 4, or 5 round stickers on one side and the words "Shop Card" printed on the reverse
- 40 yellow index cards, each with 1, 2, or 3 round stickers on one side and the words "Save Card" printed on the reverse
- Bank Statement fill-in sheets (as pictured)

Bank Statement
Spent _____
Saved _____

Description

In this game, children pretend to go shopping, but they also have to save some of their money. On each turn, players first draw a green Shop Card and take a number of coins equivalent to the number of stickers on the card. They place the coins in front of them (not on their game board). Next, they draw a yellow Save Card. From the coins in front of them, they must remove the number of coins indicated on the save card and place these coins in their piggybank or jar. The coins remaining on the table are then placed on the game board to "buy things." Play continues until one player (or both, if desired) has covered all the spaces on his or her game board. Then the accounting process begins. Children can determine how many items they bought by counting the coins on their game board. They can determine how much they saved by counting the coins in their piggybanks. The totals can be entered on their bank statements.

The Shop and Save game encourages children first to create an equivalent set by taking as many coins as there are dots on their Shop Card, and then decompose the set so that some of the money can be saved. Because the quantities are small, children can visualize the results of the subtraction process. For example, if a child takes 5 coins (the largest amount in the game) and then puts 2 coins into his piggybank, he can see that 3 coins remain. The game mimics real-life situations for many children. They realize that when a parent buys something, money must be paid. Some money usually is left over and transferred back to a pocket, purse, or wallet.

Math Discussions

This game provides many opportunities for mathematical problem solving and communication. During game play, teachers can ask questions or make comments that focus children's attention on their initial set of coins and the amount that is left when some coins are put into the piggybank. For example, the teacher might say, "How many coins did you get to take? 4? How many will you have left after you put 2 into your piggy bank? Okay, you get to buy 2 things." The teacher also can focus attention on the addition that occurs on the game board. When a child is taking her second turn, the teacher might say, "Sophie has 3 coins already on her board. I wonder how many she'll have after she adds 2 more to her board." Adults should be careful, however, to moderate their questioning. Children get tired of being asked questions and may want to focus just on the game. If children are playing in small groups, the teacher can move from game to game, adding reflective comments. The teacher also can become a participant in the game and comment on the math he or she is doing. This might be an example: "I need only 3 more coins to fill my board.

Oh boy! I have 1, 2, 3, 4 on my Shop Card. Now for my Save Card. Oh no. I have to put 2 coins in my piggybank. Look. Now I don't have enough left to fill my 3 spaces."

Interesting discussions can occur as children complete their bank statements. Following are some possible questions for them to think about:

- *Did you spend more money or save more?*
- *How much money have you saved?*
- *Do you have enough money left to fill up your game board again?*
- *Which one of you spent the most money? Who saved the most money?*

Path Games

Path games that involve quantification, such as those described in Chapter 2, are concrete representations of number lines. Children move forward along this simulated number line on each successive turn. Even though numerals typically are not used to label the spaces, children can count or visualize the total spaces that they have moved. This helps them develop the concept of moving forward by units as they pass more and more spaces. In later grades, they will be asked to apply this concept to performing arithmetic operations on a number line.

Using numerals to label the spaces on path games can be confusing to young children. It seems counterintuitive to end up on a number that does not match the quantity on the die, and it may confuse their understanding of cardinality. Children often feel that they must set their mover on the numeral designated by the current roll of the die. For this reason, if a child's mover is currently on 4, and the child then rolls a 3, the child is likely to reason that his mover should be placed on the numeral 3. The research cited earlier in this chapter documents that even though a child may be able to name numerals and represent them with objects, the concept of what a numeral really means may not be fully developed in that child's mind. Therefore, numerals can hamper children as they contemplate addition and subtraction. For these reasons, it is best to limit the use of numerals for labeling both dice and path spaces.

Because short-path games usually have only 10–12 spaces, only one die, with from 1–3 dots, is typically used. Two rolls of a 1–6 die would end the game rather quickly. In addition, recall that the short-path game is a transitional material for children who are moving from games with concrete objects to count, to more abstract path games. Smaller quantities on the die cause less confusion as children make this transition. Once children are very secure with the concept of moving along a path, a second 1–3 die can be introduced. This gives children the opportunity to combine small sets. Because they now may reach the end of the path too quickly (after just two or three turns), additional movers are needed. The goal now becomes reaching the end of the path with each mover.

Long-path games are easy to modify to incorporate addition simply by adding a second die. By the time young children are able successfully to play long-path games, they have already developed substantial number-sense concepts. First, they are at the counting level of quantification, which means that they understand cardinality. Second, they usually exhibit stable order counting, although they may still make errors when applying the one-to-one principle to counting, such as re-counting or skipping over a space on the path or a dot on the die. After many experiences playing path games, many children can subitize the quantities on a 1–6 die and quickly move the required number of spaces on the path. These children are ready for a challenge. Incorporating a second die into the game gives them larger amounts to quantify and introduces the concept of combining sets.

Teachers can expect to observe all three counting strategies applied to early addition (counting all, short-cut sum, and counting on), although the first two are the most common in preschool and kindergarten. By joining in the play of the game, teachers can model a strategy that is just above the thinking level of particular children. For example, if a child first counts the individual sets before counting them together (a count-all strategy), the teacher can model counting both sets together without counting them individually (short-cut sum). For children who apply a short-cut sum strategy, the teacher can model counting on. When presented with two dice, some children initially will count and move separately for each set. In this case, the teacher can model counting all. Although the teacher can model more advanced counting levels, children should be allowed to employ whatever strategy they are comfortable with. When they are ready to move on to a more advanced level, they will.

ACTIVITY 3.5

Feed the Birds

Materials

The following materials are needed for this game:

- Game board for each player, made from poster or tag board (6 × 22 inches)

- Twelve 1-inch-circle stickers for each game board, to form the path

- Square box labeled "Start" for each game board

- Image of a bird feeder, for the end space of each path

- 3 small novelty birds, glued to a wooden disk, for each player

- 2 dice, made from 1-inch cubes with from 1 to 3 quarter-inch dot stickers on each side

Description

In this game, children have three birds to move to their bird feeder. Because the quantities on the dice are small, some children immediately may begin to count all the dots together. Others may count and move separately according to the number of dots on each die.

Math Discussions

Several discussions are likely to arise as children play this game. First, children will need to decide what to do when a bird reaches the feeder before they have used the full amount rolled on the die. Can another bird use up the remainder? Children also may notice when their peers use different strategies to get the total. In one case, a child whose friend counted the dice separately asked, "Why do you do that? It takes too long. You could just count all the dots together." Within a few days, his friend also started to count all the dots on the dice together.

ACTIVITY 3.6

Get the Dog Some Bones

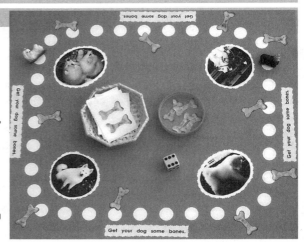

Materials

The following materials are needed
for this activity:

- Poster board, 22 × 22 inches
- 35 white 1-inch self-adhesive circles,
  to form a path around the border of
  the poster board
- Brown dog-bone cut-outs or stickers, to form bonus spaces on the
  path (see photograph)
- Basket of dog-bone cards, made from
  index cards cut in half, with from 1 to
  3 dog-bone images on each card
- Small dog mover for each player
- Bowl filled with dog-bone cut-outs, made with tag board
- Two 1–6 dot dice

Description

In order to demonstrate the wide variety of games that teachers can make related to
the same topic of interest, this game employs the dog theme used in Chapter 2. This
path game is also is a collection game. Children can move in either direction around the
path, although they may not switch directions in the middle of a turn. There is no starting or ending point. Whenever children land on a space with a dog-bone image, they
draw a card that tells them how many dog bones they can take from the bowl in the
center of the board. Older preschool and kindergarten children love collection games
such as this.

This game gives children multiple opportunities to practice adding. Each time they
take a turn, children must combine the sets on the two dice to know how many spaces to
move. As they collect dog bones, children can add the small sets together.

Math Discussions

Math talk related to this game likely will center around the number of dog bones that
each child has collected. Teachers may ask who has the most bones or how many more
a particular child needs to catch up with her friend. Another interesting topic is which
direction to move in order to reach a dog-bone bonus space. Children may calculate
whether or not they can reach a bonus space on a given turn by changing direction.

Card Games

Card games such as I Have More, from Chapter 2, can easily be adapted to support beginning addition. Two children simply draw two cards each, determine the total of the symbols on their two cards, and decide which player has more on that particular turn. That
player can take all four cards. Because the cards are being combined, the same sticker symbols should be used on all cards. It may be confusing for children to try to add apples on
one card to cats on another card. Sets should be limited to 1–3 items on each card for the
first experiences.

Large-Group Activities

Like quantification, addition and subtraction fit naturally into large-group experiences. All children should have exposure to activities that model composing and decomposing sets. Traditional songs and fingerplays often focus on adding or subtracting by 1 to model a forward or backward counting sequence. For more experienced children, these songs and activities can be altered to include addition and subtraction by numbers other than 1. The two examples that follow show how these changes might be made.

ACTIVITY 3.7

Apple Song Game

Farmer Brown had 5 red apples hanging on a tree,

Farmer Brown had 5 red apples hanging on a tree,

He plucked _____ apples, and ate them greedily,

Now Farmer Brown had ___ red apples hanging on his tree.

Materials

The following materials are needed for this game:

■ 5 felt finger-puppet apples for each child, made by cutting the apple shapes from felt and whip stitching or gluing around the edge

Description

Many teachers will recognize this traditional counting-backward song, which is sung in preschool classrooms throughout the country. In the original version, Farmer Brown picks one apple for each verse, so counting goes from 5 down to 1. In this version of the song, the teacher or a child draws a card from a small bag. The card has from 1 to 3 apple stickers on it to determine the number of apples that Farmer Brown picks. Children must figure out how many apples are left before they can sing the entire song.

Children may use their fingers to model this song, but it is easier to visualize the math (and also more fun) if they use apple finger puppets. These are easily made by cutting small apple shapes from felt and whip stitching or gluing them together. Children can put the apple shapes on their fingers, remove the required number of apples, and then subitize or count the remaining apples. Each verse begins again with 5 apples. Children thus have many opportunities to decompose 5 into its component parts. Of course, if teachers want children to decompose sets other than 5, such as 3, 4, or 6, they can start with the corresponding number of apples.

Math Discussions

Math talk occurs naturally throughout this activity as children count the apples, figure out how many to take away, and then count the remaining apples. Teachers can help by summarizing the process with such comments as, "Let's remember, now, we started with 5 apples; then we took away 3; so we had 2 left. Okay—let's sing it."

ACTIVITY 3.8

In and Out of Bed

Materials

The following materials are needed for this game:

- Literacy chart, made with poster board and sentence strips, as shown (one chart is on the flip side of the other)
- 10 teddy bears, cut from felt and mounted on poster board
- Numeral cards from 1 to 10
- Teddy bear cards, with from 1 to 3 teddy bear stickers per card and the corresponding numeral

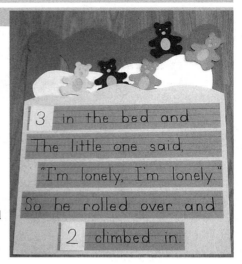

Description

Another traditional song that counts backward is "Ten in the Bed." This version uses felt teddy bears for the counters. The accompanying chart is two-sided. On one side, bears are added to the bed, whereas on the flip side, they are subtracted. As in the previous example, the number of bears to be added or subtracted is determined by a card drawn from a small bag. The card displays from 1 to 3 teddy bear stickers and the corresponding numeral. Unlike the previous activity, bears are added or subtracted until either 10 bears have been put in the bed or 10 bears have been removed from the bed. The numeral cards are attached to the chart with magnetic tape.

Bears In the Bed
☐ in the bed and
The little one said,
  I'm lonely,
  I'm lonely,
So he rolled over
and ☐ climbed in.

Bears Out of Bed
☐ in the bed and
The little one said,
  Roll over,
  Roll over,
So she rolled over
and ☐ fell out.

Math Discussions

Depending on which chart is being used, math discussions will focus on either addition or subtraction. In either case, children most likely will use counting to determine how many bears should be in the bed. For the addition chart, most children will count all the bears after a new group has been added. For the subtraction chart, they likely will count the remaining bears to determine how many are left. The teacher sometimes can model a different strategy. For example, on the addition chart, the teacher could count on from the number of bears already in the bed to get the total. On the subtraction chart, the teacher could count backward from the number of bears in bed and comment, "Let's see. There

were 7 in the bed, and now 2 fall out, so I'll take bears 7 and 6 out, and we're down to 5."
Do not be surprised if children protest when alternative strategies are introduced. It is
important to go back and model the math their way, as well, so that they can see that the
answers are still the same. Even then, children may not be convinced that the teacher's
strategy is also correct.

Operations Throughout the Curriculum

Daily living and playing experiences that occur throughout the curriculum provide many
opportunities for children to construct addition and subtraction concepts. Snack and lunch
experiences already have been discussed. Following are some examples of addition and subtraction activities that can be imbedded in other areas of the curriculum.

Dramatic Play Area

Mock restaurants, grocery stores, and ice cream parlors are often incorporated into the dramatic play area. All of these set-ups can encourage addition. As children order from menus
or purchase food labeled with price tags, they must figure out how much they need to pay.
Small numbers on the price tags, perhaps accompanied by the appropriate number of penny
stickers, support children making the transition into addition.

Block Area

Microplay, in which children use small objects to create pretend environments and stories,
encourages both addition and subtraction. In the block area, children often build fences and
enclosures when farm or zoo animals are available. The teacher can use these opportunities
to intentionally introduce addition and subtraction. For example, if a child has 3 sheep in a
pen, the teacher might ask whether 2 pigs can come and visit, and then point out that there
are now 5 animals in the pen. Conversely, the teacher might introduce subtraction by
saying, "I think this field has too many horses. Let's make another field. Now 3 horses can
eat grass here, and 3 horses are left in the other field."

Gross Motor Area

Children like to count while they perform physical actions. In this activity, children drop
two large dice (preferably of different colors) into a box or tub. The child then hops the
number of times indicated by the first die on one foot, the number for the second die on
the other foot, and the total on both feet. The teacher can help by pointing to the dots as
the child hops.

Art Area

Making collages is a popular art activity in preschool and kindergarten. Often, children have
specific objects, such as colored pasta or paper cut-outs, to glue onto their collage. Teachers
can easily introduce addition into the conversations at the art table. For example, the teacher
might say, "I see 4 red pasta wheels and 2 blue pasta wheels on your paper. Should we find
out how many that is all together?" The question can sometimes be reworded so that the
change is unknown. In this case, the teacher might say, "1, 2, 3, 4, 5, 6. I counted 6 pasta
wheels on your paper. Let's see, 1, 2, 3, 4 of them are blue, so how many are red?" This accus-
toms children to hearing addition or subtraction problems phrased in the various ways that
they will later encounter in elementary school.

Science Area

A pendulum activity is a great way to incorporate subtraction. A pendulum is a weight that is suspended by a cord so that it can swing freely. Pendulums can be suspended from a small frame such as an infant gym, or from a U-hook on a pegboard divider or shelf back. Children can use the pendulum to knock down small table blocks. Subtraction is a natural part of the process as they determine how many blocks are left to be knocked down.

UNIVERSAL DESIGN FOR SUPPORTING ARITHMETIC OPERATIONS

There is a tendency among adults to think that particular concepts are beyond the understanding of certain children, and therefore, these children should not be included in some activities. Certainly, some might think that this is true of arithmetic operations in preschool and kindergarten. Close examination of the presentation of the concepts in this chapter, however, should indicate that the activities described here are accessible at some level to all children. As an example, think about the Shop and Save game. Although addition and subtraction are the targeted concepts, the quantities involved are small. Therefore, children who are not yet counting, but who may use one-to-one correspondence to quantify, can participate. The game supports their mathematical learning at a level appropriate for them. For path games such as Get Your Dog Some Bones, a single die with from 1–3 dots per side can be substituted. This will allow children at the one-to-one correspondence level to more easily create relationships between dots on the die and steps along the path. The game also can be altered so that children hop to any point they desire, but then take only as many dog bones as the number on their die indicates.

In the suggested large-group activities, all children can participate at some level. Because the addition and subtraction are clearly modeled with objects, children with cognitive delays can participate in the counting aspects of the activities. This can provide support for the construction of the important counting principles discussed in Chapter 2. For children who are learning English as a second language, the modeling with objects that accompanies the counting allows them to understand what is happening while they learn the English words.

A multisensory approach is important for children with disabilities in the sensory areas. For children with visual disabilities, a high contrast between the background and foreground on materials is desirable. The words on cards or charts, the spaces on paths, and the objects on grid boards can be outlined with puffy paint so that children with visual disabilities can feel them. Raised dots can be used on dice. For children with hearing disabilities, the modeling with objects is particularly important. Gestures are also helpful. For example, on the Bears Out of Bed chart, when the numeral 2 is placed on the chart, the teacher might hold up 2 fingers as he points to the numeral. Children with attention-deficit concerns often find it easier to attend to singing than to talking. The songs used at group time to accompany the math concepts may be particularly helpful for them.

FOCUSING ON MATHEMATICAL PROCESSES

All of the activities presented in this chapter incorporate the five NCTM process standards, especially when the teacher adds questions or comments that encourage thinking and discussion. Next are some specific examples that illustrate the importance of these standards to the implementation of the activities.

- *Problem Solving*—Although this standard is a necessary component of every activity, it is particularly evident in examples such as Silje's figuring out how to keep score with large numbers, Jesse's calculating how many more words he needs in his word bank, and

Shawn's discovering a method for repeated addition to determine how many spoons are necessary for a cooking activity. No one told these children how to solve the problems; instead, adults supported their self-devised strategies for finding the answers.

Adult scaffolding is apparent throughout Brad's interactions with the children in his class, who are trying to model a complicated (for them) division problem. He models each of their ideas, which allows them to observe the results, and makes subtle suggestions such as "Should we count them again?" when children seem stumped. In doing so, he encourages them to keep thinking, knowing that they eventually will try something that works. Brad nurtures the children's autonomy and confidence as mathematical thinkers. These are important dispositions for later success in mathematics.

- *Reasoning and Proof*—Children who are playing mathematical games sometimes demand reasoning and proof from their peers, particularly when they think another child may be gaining an unfair advantage. More often, however, it is teachers who encourage this process. In the Eggs in the Basket activity, the teacher notices that one child's number-4 basket has 2 eggs on each side, whereas the corresponding basket of another child has 3 eggs on one side and 1 egg on the other side. In commenting on this, the teacher asks whether the children are sure that each of their baskets has the same number of eggs. This requires them to carefully examine baskets that look different but are supposed to have the same number of eggs.

  Another good example of a teacher's supporting Reasoning and Proof occurs in the Socks in the Laundry activity. Knowing that the design of the activity makes it easy for children to use one-to-one correspondence in order to compare sets, the teacher asks the children how they can tell whether one clothesline has more socks than the other. This focuses their attention on the reasoning necessary to convince their play partner about the equivalence or nonequivalence of the sets.

- *Communication*—Because the conversation that takes place during math activities is such an important factor in children's understanding of math concepts, math discussions have been included with each activity. The Shop and Save activity is one that is especially likely to stimulate substantial communication because children must repeatedly spend and save money. Card games, such as I Have More, by their very nature stimulate conversation because children must communicate about who is going to take the cards. The most extensive communication may occur during dramatic play, when children must solve math problems embedded in their play scenarios. Are there enough plates for everyone? How much did you buy? What do I have to pay for 2 scoops of ice cream? These are just a few of the mathematical situations likely to arise when the dramatic play area is planned with math outcomes in mind.

- *Connections*—Many of the scenarios in this chapter illustrate the connection between arithmetic operations and daily life. Addition and subtraction during lunch and snack, division of play materials, and scoring during gross motor play are a few examples. The connection between mathematics and literacy is highlighted when teacher and children model the division of cookies in a popular children's book. Opportunities to think about arithmetic operations occur throughout the day, and effective teachers capitalize on them.

- *Representation*—Throughout this chapter, children have been presented as representing arithmetic operations in many different ways. Jesse used his fingers as place holders for the units in a missing addend as he counted forward to determine how many more words he needed in his word bank. Shawn used finger taps to represent objects that he needed to count more than once to get the total. Children used pictorial representations of sets (e.g., dots on dice) and physical objects (e.g., coins) to calculate how many to add or subtract in games such as Shop and Save. Art was also used to illustrate mathematical operations. Young children invent and use these symbols to solve mathematical problems because they need them. Mathematics is abstract. Objects, gestures, tallies, and pictures make mathematical relationships more concrete and understandable.

ASSESSING UNDERSTANDING OF ARITHMETIC OPERATIONS

Anecdotal records and checklists can help teachers document children's mathematical development. Chapter 2 illustrated how the games and activities created as instructional tools can also be used for formative assessment. The same is true for documenting children's transition into arithmetic operations.

Children who are beginning to add and subtract have progressed beyond the initial quantification stages of global and one-to-one correspondence and now consistently use counting to quantify. For children at this level, teachers should focus on the strategies that the children use when combining (adding) or deconstructing (subtracting) sets. As the following examples show, this information can help teachers decide what they should do to support learning:

EXAMPLE 3.8

Nina and Maya are excited when the teacher gives them a second die to use while playing a path game. Nina takes the first turn. She carefully counts four dots on the first die and five dots on the second die. Then she starts over and counts nine dots before moving nine spaces along the path. Maya then rolls the dice. Maya counts three dots on her first die and moves three spaces. Then she counts five dots on her second die and moves five more spaces. "I'm right behind you," Maya tells Nina.

In this example, both girls are making the transition into addition; however, they are at different levels of understanding. Nina realizes that she can get the total by counting all of the dots on both dice together, but she first needs to count each set separately. Nina is at the counting-all level of early addition. Maya has not yet reached this point. She counts each of her sets separately before moving. The teacher was wise to give these girls a second die to use while playing the game. Through experience, Nina will realize that she does not really need to count the dice separately before combining them. If the teacher joins the game, she can model the shortcut-sum strategy of immediately counting all of the dots together. By observing her friend Nina, Maya likely soon will realize that she too can combine both dice without counting them separately.

EXAMPLE 3.9

Barry rolls two dice. "I got 2 and 2," he announces before moving four spaces. On his next turn, Barry again subitizes the quantities on the dice. "I got 3 and 5," he says. Then he counts the eight dots.

Barry's quick announcement on his first turn shows that he knows the addition combination "two plus two." On his second turn, he uses a shortcut-sum strategy to determine the total. Because Barry can subitize the sets of dots on the dice and clearly understands what they represent, the teacher can model by using a counting-on strategy when it is her turn. This probably will cause Barry to think hard about the fact that he already knows how many dots are on the individual dice and may not need to re-count them.

Table 3.1. Addition checklist

Child	Counts sets separately	Counts all	Shortcut sum	Counts on	Knows combinations
Nina		X			
Maya	X				
Barry			X		2 + 2

Although anecdotal notes are quite useful to teachers, sometimes a quicker method of documentation is desirable. A carefully constructed checklist can capture the important information and takes just seconds to complete. Table 3.1 is an example of a chart that teachers might create to document progress in addition. It illustrates how the information in the anecdotes about Nina, Maya, and Barry might be recorded.

SUMMARY

This chapter extends educators' understanding of children's development of quantification concepts to include children's transition into addition. At first, children need to count each set individually before counting the items in both sets together. This is called a count-all strategy. When children have had many opportunities to combine sets, such as by rolling two dice while playing quantification games, they realize that they can immediately count all of the objects (e.g., dots) together. This often is referred to as a shortcut sum strategy. Children usually remain at this stage for a considerable period. Finally, children realize that the individual sets they are combining are parts of a larger whole. At this point, children count on from one of the sets. Although older children and adults typically count forward from the larger of two sets, younger children may prefer to count forward from the smaller set.

After repeated experiences in which they combine sets, children begin to remember addition combinations. The easiest combinations to remember are the doubles up to $5 + 5$, perhaps because children so easily can represent them by counting the same number of fingers on each hand. Next, children become fluent at adding $+1$ and then $+2$. They build on this knowledge to solve addition problems involving other combinations. For example, a child might say that $2 + 3$ is 5 because she knows that $2 + 2$ is 4, and 3 is 1 more than 2.

Through their play, young children engage in all four arithmetic operations. For subtraction, children may use fingers or other objects to represent the starting number, and model lowering fingers or removing objects as they subtract. They may also use a counting-on strategy to compute the number of units in between the set they are subtracting and the total amount. Children may use taps or other gestures to represent repeated addition (i.e., multiplication). Division strategies follow a developmental progression. At first, young children use a global strategy of giving each person an amount that looks about the same. This is followed by the use of one-to-one correspondence, in which each individual is given one item at a time until all items have been distributed, much like dealing cards. Older children begin to decompose the units in a set to determine how many each person should receive. They also may employ repeated subtraction. For example, each person might be given two items, and if there are still some left, then one more each.

Teachers should capitalize on opportunities throughout the day to help children think about arithmetic operations. Snack time provides endless opportunities to model all four operations. Carefully planned small- and large-group experiences support children's construction of these beginning arithmetic concepts and gives them the opportunity to explore the relationships among operations. Assessment should be ongoing and an integral part of these experiences.

ON YOUR OWN

■ Reflect upon the daily activities of preschool and kindergarten children. What arithmetic operations could be incorporated into these activities?

■ Modify an existing number-sense game or activity (see Chapter 2) to incorporate addition or subtraction.

■ Think of some ways to illustrate for children the connection between addition and subtraction.

REFERENCES

Baroody, A.J., & Tiilikainen, S.H. (2003). Two perspectives on addition development. In A.J. Baroody, & A. Dowker (Eds.), *The development of arithmetic concepts and skills* (pp. 75–125). Mahwah, NJ: Erlbaum.

Carpenter, T.P., Fennema, E., Franke, M.L., Levi, L., & Empson, S.B. (1999). *Children's mathematics: Cognitively guided instruction.* Portsmouth, NH: Heinemann.

Clements, D.H., & Sarama, J. (2007). Early mathematics learning. In F.K. Lester, Jr. (Ed.), *Second handbook of research on mathematics teaching and learning* (pp. 461–555). Reston, VA: National Council of Teachers of Mathematics.

Fennell, F. (2010, April). *Focus on Fractions, Building Fraction Sense: Why Not?* Presentation at the NCTM National Conference, San Diego.

Hutchins, P. *The doorbell rang.* New York: Greenwillow.

Kamii, C. (2000). *Young children reinvent arithmetic: Implications of Piaget's theory* (2nd ed.). New York: Teachers College Press.

Kato, Y., Kamii, C., Ozaki, K., & Nagahiro, M. (2002). Young children's representations of groups of objects: The relationship between abstraction and representation. *Journal for Research in Mathematics Education, 33*(1), 30–46.

Moomaw, S., & Hieronymus, B. (1995). *More than counting.* St. Paul, MN: Redleaf Press.

Moomaw, S., & Hieronymus, B. (1999). *Much more than counting.* St. Paul, MN: Redleaf Press.

National Council of Teachers of Mathematics. (2000). *Curriculum and evaluation standards for school mathematics.* Reston, VA: Author.

4

Developing Algebra Concepts

My mom had a huge button drawer. Her father had been in the rag trade, and so there was an endless supply of buttons of all kinds. I spent hours playing with them, sorting them into categories, some obvious, such as size, color, and shape, and some I made up. Sometimes I sat at the coffee table, meticulously organizing, carefully sorting. Other times I lay on the floor in my bedroom, buttons at eye level, seeing this ever-moving kaleidoscope of my own making, often seeming to take on a life of its own.

—Irene M. Pepperberg (2008, p. 34)

Algebra, a term that fills so many with fear, has its roots in the play of young children. Just as Irene Pepperberg did as a child (she is now an animal cognition researcher at Brandeis University), children gather objects of interest and incorporate them into their play. Pebbles may be organized into categories according to color or size and sorted into empty bottles. Buttons may be arranged in a symmetrical **pattern.** Birthday cards may be lined up in a sequence from most to least favorite. All of these childhood activities contain algebraic ideas, yet children embrace them with wonder, intensity of thought, and contentment. School mathematics should be no less inviting.

THE ALGEBRA STANDARD

The National Council of Teachers of Mathematics (NCTM) views Algebra as a curricular strand that begins in preschool and continues beyond Grade 12. As with numerical concepts, the foundations of Algebra develop during the early years; so, the preschool and kindergarten curricula are important. Adults often equate algebra with the manipulation of obscure symbols; however, it is the understanding of the concepts that these symbols have been created to represent, along with the mathematical principles that govern their use, which is critical. When viewed in this manner, it is easy to trace the beginnings of algebra back to early childhood.

The Algebra standard (NCTM, 2000) encompasses the following concepts:

- Understanding patterns, relations, and functions

- Using algebraic symbols to represent and analyze mathematical situations

- Using models to represent quantitative relationships

- Analyzing change

Remember that these standards represent a continuum from preschool through high school, and development will look very different at each point along this continuum. So, what do these concepts mean in the context of preschool and kindergarten classrooms?

Patterns can be thought of as relationships involving a repeating element. Our number system is based on patterns: Each whole number is one more than the previous number; each place-value column (moving from right to left) is 10 times the unit value of the column to its right; every even number is followed by an odd number; and so forth. Patterns provide a framework for interpreting and understanding mathematics.

Before children can begin to organize objects (much less, numbers) into patterns, they first must be able to focus on particular aspects of the items. On the basis of these traits, children can sort objects into groups (e.g., red, yellow, or green), develop classifications (e.g., farm, zoo, and aquarium animals), and put the items into an ordered system (e.g., smallest to largest). Pattern relationships develop as another way to organize the items. Therefore, sorting and classifying, along with patterning, are important elements of the Algebra standard in preschool and kindergarten. Eventually, as children move through elementary and middle school, they apply the logic that they have developed through creating relationships among concrete objects to creating numerical relationships. For example, they sort numbers into whole numbers and fractions, develop a place-value hierarchy, and recognize numeric patterns, such as the alternation of even and odd numbers when counting.

Obviously, young children do not use standard mathematical notation. This understanding develops gradually over time. Children do, however, invent symbols to represent mathematical situations. They may use tally marks to keep track of a score, fingers to communicate how many candles are on a birthday cake, or hand claps to represent rhythmic patterns in music. These symbols help them think about mathematical implications. For example, in a vignette in Chapter 2, the days of kindergarten were represented by teddy bear tags added each day to the bulletin board. By watching the teacher count these tags in two directions over the course of several weeks, children eventually constructed the order-irrelevant counting principle. Although this discovery happened in the number-sense realm, the generalization that occurred is an algebraic structure.

Closely related to representation is the ability to model a mathematical problem. Children may use manipulative materials, drawings, and actions to model problems. Children also notice change. When they roll cars down a ramp, they observe that the cars roll faster when the ramp is steeper. They notice that plants get taller as they grow. Eventually, they will be able to use mathematics to describe these changes and make predictions.

The scenarios that follow illustrate ways in which algebraic relationships may appear in the play and discussions of young children.

EXAMPLE 4.1

Tamara's kindergarten class was fascinated with dinosaurs. As part of her integrated dinosaur curriculum, Tamara assembled a collection of small dinosaurs for the children to sort and classify. She expected that they might sort the dinosaurs into categories on the basis of color, size, or type of dinosaur. Some of the class "experts" might even classify them as plant or meat eaters.

Tamara observed as 5-year-old Jeff played for a long time with the dinosaurs. He carefully shifted dinosaurs among the various piles that he created. When it appeared that Jeff had finished, Tamara approached to take a closer look. It was immediately apparent that there were three dinosaurs in each pile; beyond that, there seemed to be no noticeable attribute that distinguished one group from another.

"What were you thinking about when you put these dinosaurs into groups?" Tamara asked.

"Well," said Jeff, "I wanted three dinosaurs in each group, and I wanted all three dinosaurs to be different from one another."

"Oh my," Tamara thought to herself. "I thought he couldn't sort the dinosaurs by any characteristics, and he created his own 'all different' attribute to sort them by."

This example illustrates the amount of thought that young children may put into organizing interesting materials. When they sort objects into categories, they usually think about some way in which various items are alike. Color, size, and type of material are typical attributes that children might use to group items. In this example, however, Jeff focused on the concept of *different* and created sets in which no two dinosaurs were alike. Simultaneously, he considered the number in each category and placed exactly three different dinosaurs in each group.

Much later, when children such as Jeff formally study algebra, they will be asked to consider mathematical concepts such as equalities and inequalities. Equal, unequal, alike, different, similar, more, and less are some of the many mathematical relationships that children explore as they sort and classify collections of objects. The roots of algebra, therefore, extend all the way back to preschool and kindergarten and the relationships children construct through their play.

EXAMPLE 4.2

Tuffa's preschool class was getting ready to go outside. "Go to your cubbies and put on your jackets," the teacher announced. Tuffa started toward his cubby, but then noticed the can band in the music area of the classroom. There were five cans, each a different size, and a mallet for playing them. Tuffa picked up the mallet and began playing the cans in a random order. He noticed the different sounds that each can produced when he hit the top with his mallet. Soon Tuffa's playing evolved into a complex pattern that he repeated over and over. A visitor in the class observation booth listened in awe to the 4-year-old child's pattern. The adults in the classroom, however, did not appear to notice, as their attention remained fixed on the transition to outside.

Young children live in a world filled with patterns: the checkered design of a tiled floor; Grandma's doorbell chimes, which play the same tune each time the button is pressed; the step-hop movement on alternating feet used for skipping; the back-and-forth movement of a swing. With so many sources of patterns available to explore with young children, it is unfortunate when teachers limit their patterning curriculum to wooden cubes placed in an alternating color sequence or shapes that the teacher arranges in a pattern on a daily calendar.

As this vignette illustrates, children may create elaborate patterns when presented with a medium that resonates with them; however, these patterns may go unnoticed if teachers focus only on the more typical patterning materials. The creation of patterning relationships is another early indicator of algebraic thinking.

EXAMPLE 4.3

Several preschool children were busy drawing pictures at the art table. Suddenly, Nancy turned to her teacher and asked, "How can you tell if a number is even?" Astounded, the teacher was momentarily speechless while she tried to decide how to explain the concept of even numbers to a 4-year-old. She need not have worried. "I know," Claire, another 4-year-old, offered. "A number is even if two friends can share and each has the same amount."

Young children's knowledge can be surprising. Where did Nancy hear the term *even number*, and how did Claire come up with such an astute definition? Nancy may have heard her older siblings talking about even and odd numbers, and she may have wondered about it. It is likely that Claire came up with her definition after many experiences with sharing. When items are divided between two friends, children often deal them out, one by one, to each individual. When they are finished, they may say, "There, that's even." It may be through experiences such as these that Claire decided that having an even number must mean that each of two people has the same amount. The scenario shows the intricate mathematical relationships that children develop through their concrete experiences with number concepts. Claire is already forming some overarching concepts—that is, *even* can apply to any number that can be shared equally between two people. Like sorting, classifying and patterning, the development of big mathematical ideas is also evidence of the beginning of algebraic reasoning.

EXAMPLE 4.4

Maryam, a child care provider, was caring for four toddlers in her home. She had often noticed that these 2-year-olds liked to place felt cut-outs onto a flannel blanket that she kept draped over the back of the sofa. Building on this interest, Maryam decided to create some new felt cut-outs that the children might pair up, or put into a one-to-one correspondence relationship. She traced and cut out shapes of children and umbrellas and placed them in a basket on the sofa.

 The next day, Andy was the first child to be dropped off at child care. "Look, Andy," said Maryam, pointing to the new felt cut-outs. "There are people and umbrellas. Can each person have an umbrella?" Maryam carefully placed an umbrella cut-out above one of the people. Delighted, Andy began lining up the people. Then he placed an umbrella above each person. Together, he and Maryam sang, "Rain, Rain, Go Away."

One-to-one correspondence is a foundational mathematical concept. Children build on this concept in early quantification when they align objects to determine "more," "less," or "the same." The concept is further extended when children begin to apply one counting word to each object that they are quantifying. Young children like to put things into a paired relationship. They may set one toy person on each block, put a hat on each doll baby, or place a toy cup on each saucer. Because many toddlers have had the delightful experience of walking under an umbrella in the rain, they realize that umbrellas and people go together. Maryam builds on this "go-together" relationship with her felt cut-outs, which encourages the toddlers to put two sets (people and umbrellas) into a one-to-one correspondence relationship. The one-to-one relationship is yet another big mathematical idea that forms the roots for later algebraic reasoning.

THE DEVELOPMENT OF ALGEBRAIC REASONING

Considerably less research has been conducted regarding the development of foundational algebraic concepts in young children than has been conducted regarding the development of concepts of number and operations. *The Second Handbook of Research on Mathematics Teaching and Learning* (Lester, 2007) devotes less than one page to algebraic thinking in early childhood, whereas there are 22 pages summarizing research on number and operations and 29 pages on geometry and spatial reasoning. Nevertheless, the National Mathematics Advisory Panel (U.S. Department of Education, 2008) places a strong emphasis on the preparation of all students for algebra.

If the research community has been relatively silent on the development of algebraic thinking, the mathematics education community has been more forthcoming. As previously discussed, NCTM has developed learning outcomes for foundational algebraic concepts that extend down to the preschool years.

Sorting and Classifying

Kindergarten teacher and author Mary Baratta-Lornton (1976) describes sorting and classifying activities as critical for young children because they encourage analytic thinking and clear communication. Deciding upon groupings, and determining the relationships within and among groups, helps children develop logical reasoning. Kamii (1982) also advises teachers to encourage children to put all kinds of objects and actions into all kinds of relationships because this develops autonomous thinkers.

When children first begin sorting objects into categories, they often are able to focus on only a single attribute, such as color. They may want to group bottle caps or buttons always on the basis of this attribute. For other collections, such as toy animals or doll shoes, the predominant sorting element of choice may be type. Because one of the purposes of sorting and classifying activities is to help children think divergently, the teacher's role is to support multiple ideas for sorting criteria. For example, the teacher might suggest that all of the bottle caps with flip tops be placed in one section of a sorting tray. This action results in the grouping of many colors of bottle caps within the category of "flip tops." The following types of questions and comments may help children learn to focus on a variety of attributes when sorting and classifying items:

- *Is there any other way to sort these pom-poms? Some look really big, and some are very small.*

- *Do all of the buttons have the same number of holes? I'd like to find all of the buttons with four holes. Can you help me?*

- *What should we do with all of these shells? Can you help me sort them?*

- *Look at the group I made with the keys. What is the same about all of the keys?*

- *Can I put this paperclip in this bowl? Why not?*

- *What should I do with this bandage? It has red and white stripes. Does it go with the red bandages or the white ones?*

The last question in the preceding list points to an interesting problem that develops as children sort open-ended collections. What do you do when an item can fit into more than one group? Baratta-Lornton (1976) emphasizes that it is important for children to determine how to handle these problems. Some may decide that the striped bandage cannot go into either group; others may want to put it in between the red and white groups. At this point, the teacher might introduce overlapping yarn loops and talk about the possibility of being in both groups at the same time. Similar classroom situations could be discussed. For example, Simone might be in the "girl" group, but also in the "black hair" group, or the teacher might sometimes position her chair so that she is in both the manipulative and book areas.

An important part of the sorting process is children's communication about how they make their decisions. Teachers can guide them to be more specific in their descriptions. For example, if a child says that the items in a particular group are the same, the teacher might ask, "How are they the same?" In addition to helping children express why an object fits in a particular group, it is also important to focus on why an object *does not* fit in that group. Expressing a negative application is often more difficult than justifying a positive one.

To summarize, teachers should expect young children to first focus on only one possible attribute for sorting a collection of objects. With repeated experiences, children begin to focus on other attributes to use as the criteria for categorizing objects. This transition in thinking will likely happen more quickly if another child or the teacher introduces other ideas for sorting the same items. Eventually, children will realize that an object can be part of more than one set. They may invent a way of showing this intersection, or they may describe their rationale.

Patterning

Another foundational component of early algebraic thinking is pattern recognition (NCTM, 2000). Recognizing and extending patterns can be challenging for young children because they must perceive individual elements in relationship to the whole. There is little research to clarify how young children construct patterning concepts; however, one study conducted in Australia indicated that 76% of children entering kindergarten could copy a repeating color pattern, but only 31% could explain or extend it (Clements & Sarama, 2007). Repeating a color pattern could simply involve matching, so this would not indicate a true understanding of the concept of a pattern. Recent research showed that, in a sample of 105 preschool children, 60 (57%) could extend a pattern involving two alternating colors and 39 (37%) could extend a pattern involving three repeating colors (Moomaw, 2008).

Patterns are logical–mathematical relationships that must be constructed by the individual. Descriptions and explanations given by the teacher, no matter how clearly expressed, often do not help. Children must form these relationships themselves, and teachers can only provide the opportunities for this to happen, as the following instructive episode of mine illustrates.

EXAMPLE 4.5

One of my biggest "aha" moments as a teacher came after I had tried unsuccessfully for a week to introduce the concept of patterning to a young kindergarten class. I had created pattern strips with alternating images of a jack-o-lantern and a black cat. I hoped that the children would recognize the pattern and use cut-out images of jack-o-lanterns and cats to extend it. No matter how many times I explained the activity, they just did not get it. I decided that they were not developmentally ready for the concept of extending a pattern.

EXAMPLE 4.5 *(continued)*

Because my undergraduate background was in music education, I always included music as a daily activity. An important component of these experiences with music involved clapping rhythmic patterns or playing them on instruments. Because my jack-o-lanterns and black cats had failed as a patterning activity, I decided to recycle them into a rhythm activity. I pointed to the pictures while I rhythmically chanted jack-o-lan-tern CAT, jack-o-lan-tern CAT. Soon, the children were clapping the pattern and chanting along with me. Then an amazing thing happened. As we reached the end of the pattern strip, the children continued chanting the pattern. Laughing at my shocked expression, they extended the pattern on and on.

Somewhat dubiously, I then picked up a different pattern strip and started a new chant: CAT CAT jack-o-lan-tern, CAT CAT jack-o-lan-tern. The class joined in and, once again, extended the pattern when we ran out of images. From that point on, the class eagerly pointed out patterns throughout the classroom, created their own patterns, and extended the images on my pattern strips. In subsequent years, I introduced patterning to preschool classes with equal success by using the same method of chanting, clapping, and moving to the rhythm of the chanted patterns. In talking with preschool and kindergarten teachers at seminars throughout the country, I have found that many, by happenstance, have made the same discoveries that I did. If you start by singing or chanting a pattern, children quickly pick up the idea of repetition. They soon make the transition to recognizing visual patterns.

The easiest patterns for children to recognize and extend are alternating patterns in which the items vary by only one attribute, such as color or shape (Moomaw & Hieronymus, 1999). These are often referred to as AB patterns. An example would be stringing beads in a red-blue pattern. It is difficult for young children to focus on more than one aspect of a pattern at a time. Therefore, the pattern "red circle–blue circle, red square–blue square" would be more difficult than simply alternating red and blue circles. Teachers often observe that some combinations of repeating elements are confounding for children; for example, the pattern "green-yellow-green, green-yellow-green" is confusing because children notice the two green items next to each other and want to end the pattern with two greens. Chanting is particularly useful in this type of situation because it helps children decompose the pattern into groups of "green-yellow-green." The chant would be as follows: green-yellow-green (pause), green-yellow-green (pause).

Algebraic Representation

Another foundational concept of algebraic reasoning is the understanding that one thing can represent another. Toddlers begin to develop this concept when they use a block to represent a comb one minute and a car the next. During preschool, children learn that a particular configuration of letters represents their name—or them. They also learn that a series of dots on a die can represent a particular quantity, and so can a numeral. As they begin to understand cardinality, children realize that when they count a set of objects, the number word given to the last object represents the entire set. When playing games, they invent hash marks or hold up fingers to represent their score.

All of these examples illustrate the ways in which algebra intersects with other areas of mathematics from the earliest years. By building on children's natural inclinations to let an object or symbol represent something else, teachers can help children model mathematical problems so that they can reason more clearly about the underlying concepts. Suppose that kindergarten children are presented with a scenario in which there are seven fish in a pool, but four fish swim away. Children are happy to use cubes to represent the fish. The cubes allow them to visualize seven "somethings" and watch the transformation when four are taken away. As they progress in mathematics, children will learn many ways to use symbols to represent mathematical situations. Nevertheless, the foundational concept remains the same. They can use one thing to represent another.

DESIGNING THE ALGEBRA CURRICULUM

Algebra is interwoven throughout the number sense and operations curriculum discussed in Chapters 2 and 3. As children attempt to answer important numerical questions such as "Who has more?" they must differentiate between the groups that they are comparing and create a mathematical model, perhaps by aligning items in the two sets. These are components of early algebraic reasoning. The very structure of our number system is based on patterns, another component of algebra. Algebra permeates the Geometry, Measurement, and Data Analysis and Probability content areas as well.

Although it is important to recognize the connections between algebra and other areas of mathematics, teachers can also design a curriculum that highlights, and helps children construct, the fundamental algebraic concepts of sorting and classifying, creating patterning relationships, and modeling mathematical problems. Once again, it is helpful to think of the curriculum in terms of math talk, individual or small-group activities, large-group activities, and curriculum interwoven throughout all areas of the classroom.

Math Talk

There are endless opportunities throughout the day to encourage mathematical reasoning through conversations with children. Chapter 3 discussed snack and lunch routines that provided the catalyst for mathematically centered conversations. The example that follows shows the way in which a teacher's timely comments may stimulate modeling of a mathematical situation—an important component of algebra—through a child's individual play.

EXAMPLE 4.6

Three-year-old Juan was intently building a train with interlocking blocks with wheels. He had hooked four cars onto an engine when the teacher sat down beside him. "I see you've made a train," the teacher said. "There are some people in this basket," she continued, pointing to a nearby container of Duplo toy people. "I wonder what your train would look like if each car had a passenger."

Juan seemed very interested in this possibility. Reaching into the basket of people, he carefully hooked one person onto each car in a one-to-one correspondence relationship. The teacher counted the people and said, "Now you have four passengers on your train." Juan looked pleased.

"Could each car have two passengers?" the teacher now asked. In response, Juan added another person to each car. Together, he and the teacher counted eight passengers.

Building materials provide perfect opportunities for teachers to integrate mathematical representation into children's play. Teachers might, for example, focus on size comparisons by asking children to estimate how many horses, versus how many sheep, will fit into the barn that they have built. On another occasion, teachers could draw attention to patterning relationships. Many children build symmetrical block structures, and this could be discussed with the class architects. The teacher could even put in an order such as, "I would like to have you build a store for me. I want a pattern of triangle, triangle, square to go around the roof. Can you draw up some plans?" This conversation encourages children to represent the problem in two ways, first by drawing a model and then by building it.

To return to the scenario with Juan, on subsequent occasions, the teacher might encourage other representations. The list that follows provides some examples.

- *Here are some animals that you can also use for your train. I don't think the people want to ride with the animals, though. Could you put the people in some cars and the animals in other cars?*

- *I see people in these cars and animals in these cars. Are there more cars with animals or people?*

- *Today we have zoo animals. What if each zoo animal needs an animal trainer to ride in the car with it? What would that look like?*

- *Can we make a pattern with people facing forward and people facing backward?*

When teachers add particular materials to the classroom, they should speculate on the opportunities for mathematical modeling that the materials present. Teachers should then talk with their staff about the types of mathematical conversations they can have with children who are using the materials.

Individual or Small-Group Activities

Sorting and classifying collections of objects, as recounted in Irene Peppercorn's recollections at the beginning of this chapter, is an activity that can be quite intriguing to young children. Each day they can impose a new structure on the same group of items. The same is true of patterning activities, once children begin to realize the endless creative possibilities. Both types of experience are also enjoyable to share with a friend or a small group of children, who may work together to impose order on a collection or to create patterns for one another to extend. The third type of algebraic experience that teachers can design for individuals or small groups is the modeling of mathematical problems and relationships.

Collections for Sorting and Classifying

Collections are groups of objects which share observable attributes that children can use to sort the items into groups. Color, shape, texture, material, size, and design are some of the many attributes that children may focus on as they categorize materials. From a cognitive-learning standpoint, the best collections are those in which the objects share several attributes so that they can be grouped in more than one way. This helps children create different relationships among the materials and begin to think more flexibly. Consider, for example, a collection of seashells. The shells may vary by color, such as pink, brown, white, or multicolor. Within these various colors, there may be large, medium, and small shells. The texture may vary from shells that are smooth to those with ridges or bumps. Shapes may include spiral, scallop, and asymmetrical, whereas designs could vary from striped to dotted to variegated. Because the shells share so many different attributes, a given shell might be grouped with the spotted shells one day, the medium-sized shells another day, and the spiral shells on a third day.

Collections that teachers develop are usually more interesting and educationally useful than those which are commercially available, because they have more attributes. An example of a popular commercial material is teddy bear counters. Although they are useful as counters or movers for math games, they are less appropriate for sorting and classification activities (for which they are also marketed), because they vary by only color, or in some cases, size. Once children have grouped the bears by color or size, there is no other way to classify them; however, teachers can expand the teddy bear collection and make it a much more interesting mathematical material, as in the "Better Teddy Bear Collection" described next. Ideas for teacher-assembled collections include decorative rocks and pebbles, assorted bottle caps, small novelty animals, barrettes and hair clips, buttons, costume jewelry, doll shoes, colored pasta, and children's bandages. For many more ideas about creating collections for sorting and classifying, see Moomaw and Hieronymus (1995).

ACTIVITY 4.1

A Better Teddy Bear Collection

Materials

The following materials are needed for this activity:

- Plastic teddy bear counters (several of each color)

- Small flocked teddy bears, available in craft and party stores in a variety of colors

- Small plush teddy bears, also available in craft and party stores

- Flat, wooden teddy bears, painted in colors that match the other teddy bears in the collection (also available in craft stores)

- Teddy bears cut from fun foam, in a variety of colors (bear-shaped hole punches are available in several sizes in scrapbooking departments)

- Divided plastic tray, for sorting the bears

Description

This collection starts with colored teddy bear counters, which are commonly used in preschool and kindergarten classrooms. The collection is expanded to include teddy bears of different sizes that are made from a variety of materials. The addition of these bears provides multiple attributes that children can use to sort the collection.

Math Discussions

For collection activities, math discussions will likely begin with a recounting of how the children decided which bears should go together. Remember that, in addition to talking about characteristics that are the same, the discussion should also include characteristics that are different. A key concept in sorting and classifying is that a particular attribute is used to define which things can go together. Teachers should focus on this concept in their discussions. The teacher might ask, "Why does this bear go with this group when it so much bigger than the other bears?" This requires children to consider the general rule that they have followed in sorting the bears, while also accounting for peripheral characteristics.

Teachers can also use discussions to introduce other concepts, such as sorting by a different attribute or even thinking about two attributes. An example might be encouraging children to find bears that are red but also fuzzy.

ACTIVITY 4.2

Bandage Collection

Materials

The following materials are needed for this activity:

- Bandages that vary by size (small, medium, large)
- Bandages that vary by shape (rectangles, circles, hearts)
- Bandages of various colors (red, blue, pink, brown, and so forth)
- Bandages with designs (stripes, dots, pictures)
- Sorting box with divided sections
- Self-laminating adhesive film

Description

Children love bandages, which cover up those nasty scrapes and cuts (real or imagined). Today, children's bandages are available in an array of colors, designs, and shapes. There are even heart-shaped bandages. For this collection, the bandages are covered on both sides with self-laminating film. Simply press the bandages onto one sheet of film, cover them with another sheet, and cut them out. In this way, the bandages are durable enough for repeated handling, and children cannot "wear" them on their skin.

Parents are a good source for increasing the variety of a bandage collection. They may be willing to donate a few of their child's favorites. In one school, several teachers each bought one box of bandages, each of a different variety. The teachers then shared them so that each teacher had a large variety of bandages for very little cost. The children loved the bandage collections. They sorted them in many different ways, including the ones they liked the best!

Math Discussions

Young children love guessing games, and this interest can be incorporated into math discussions. Instead of starting the conversation with a question regarding how the items are grouped, the teacher might decide to guess. He might say, "Let's play a game. I'll close my eyes, and you give me one of the bandages. Then I'll try to figure out which group it belongs in." The teacher can then model possibilities. "My bandage is big, and I see a big bandage over here, so maybe it should go there. Is that right?" The game encourages children to explain why the teacher's logic is flawed. For example, they might say, "Your bandage is big, but it's not the right color for that group." These types of discussions help children think more flexibly about the many sorting possibilities that exist within a collection, as well as how to focus on the key attribute used to establish the categories in any given arrangement.

Patterning Activities

The collections of materials that teachers develop for sorting and classifying activities can be adapted to create interesting patterning activities for individuals or small groups of children. Items should be selected from the collections that strongly exhibit only the attributes required for the pattern. For example, if a teacher wants children to extend and create patterns with spiral and scallop shells, then all of the shells should be about the same size and color. In this way, children can focus on the shape component of the pattern and not be distracted by confounding elements.

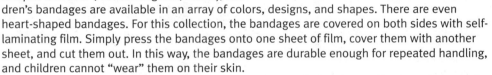

Children should be encouraged to generalize patterning concepts by representing the same pattern through various media. When using shells, children could be encouraged to copy the patterns that they create with real shells by using materials in the art area. Shell stickers, pictures of shells downloaded from the Internet, or stamps that create prints of the two types of shells are possibilities. From the child's standpoint, the art representation preserves the pattern that they originally created with the actual shells. This ability to generalize concepts and represent them in various ways is a strong component of algebra. For many examples of teacher-designed patterning activities, see Baratta-Lornton (1976) or Moomaw and Hieronymus (1999).

ACTIVITY 4.3

Going Fishing

Materials

The following materials are needed for this activity:

- Pegboard divider with hooks; or a towel rack mounted to the wall, and S-hooks
- Fish and starfish shapes, cut from colored fun foam, with a hole punched near the top so that they can be hung on the hooks
- Basket, to hold the fish
- White poster board strips, with fish and starfish pattern starters drawn on them (optional)

Description

Many young children like to pretend that they are fishing. This patterning activity ties in with that popular play theme. Children can hang the fish and starfish from the hooks to create patterns. Teachers might model this activity first at group time so that children have an idea of how they can use the materials. During choice time, children can create patterns on their own or in cooperation with other children. To help get them started, pattern strips can be hung above the hooks provided for the fish.

For initial patterning experiences, limit the patterning attributes to either color or shape; for example, include fish and starfish that are all the same color, or remove the starfish and use just two colors of fish. As children become more experienced at creating patterns, fish and starfish of various colors can be added to the collection of materials.

This activity can be incorporated into many different areas of the classroom. If the dramatic play area is set up as a campground or fishing site, the patterning activity can be integrated into that area. Children might also like to play with the fish in the water table. If a board is stretched across the top of the table, children can lay the fish on the board to create patterns. The activity can also be placed in the math or manipulative areas of the classroom.

Math Discussions

Children may simply hang the fish from the hooks until the teacher introduces the concept of patterning. Skillful teachers may participate in play situations and introduce the idea of creating patterns with the materials. For example, the teacher might say, "I caught some fish to hang up until dinner. Look, the way I hung them on the hooks makes a pattern: green fish–purple fish, green fish–purple fish."

Teachers can also help children analyze their own patterns or those of their friends. Often, young children will start a pattern but will lose sight of it after several repetitions. When the teacher chants the pattern along with the children, they often discover their errors because the word that comes next in the chant does not match what they have represented. Learning to analyze and self correct is an important skill in mathematics.

ACTIVITY 4.4

Patterning with Apples and Pumpkins

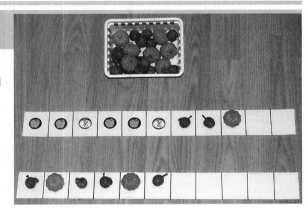

Materials

The following materials are needed for this activity:

- Pattern strips, made from poster board $2\frac{1}{2} \times 22$ inches, with vertical lines drawn every 2 inches to divide the strips into boxes
- Basket of novelty pumpkins (about $1\frac{1}{2}$ inches in diameter)
- Basket of novelty apples (about $1\frac{1}{2}$ inches in diameter)

Description

This is another type of activity that encourages children to create and extend patterns. The pattern strips invite children to place the pumpkins and apples in the boxes, and with two types of objects available, alternating patterns often emerge. Teachers may want to incorporate this activity into group-time experiences. The teacher can create a pattern, and with the children's help, chant the pattern rhythmically. In this way, a variety of increasingly complex patterns can be introduced. Later, children can reinvent these patterns on their own. If desired, stickers of pumpkins and apples can be used as pattern starters at the beginning of the pattern strips.

Math Discussions

As teachers move around the classroom, it is important that they engage children in discussions about the patterns the children create. Communication, one of the process standards, requires children to reflect on their actions and explain clearly what they have done. This skill is somewhat different from simply creating the pattern. As in the previous activity, teachers may also encourage children to chant their patterns so that they can check for errors.

Math discussions also provide the teacher with the opportunity to present other ideas for patterning. For example, the teacher might say, "I see one apple after every pumpkin. What would your pattern look like if there were two apples after every pumpkin?" Children can then translate this suggestion into a new pattern.

Teachers may want to keep available paper pattern strips, similar to the cardboard ones used in the activity, for children to use for copying the patterns that they have created. The paper strips require the child to model the pattern through a different medium and also provide a way for them to preserve their patterns.

Modeling Mathematical Situations

Chapters 2 and 3 provided many examples of children modeling mathematical problems. The children used a variety of manipulative materials to model the quantities rolled on dice; their bodies, to hop a given number of spaces along a path; fingers, to represent words still needed in a word bank; plastic eggs, to decompose numbers; and so forth. This section focuses on arrangements of materials that encourage representation and modeling of mathematical situations.

ACTIVITY 4.5

Hats and Mittens— A Beginning Function

Materials

The following materials are needed for this activity:

- Several baby dolls
- Selection of infant hats and mittens

Description

The dramatic play area would be a logical place for this activity, although the dolls and accessories could also be placed in a basket in the manipulative area. The materials support the modeling of a variety of mathematical concepts:

- One-to-one correspondence (one hat for each head, one mitten for each hand)
- Quantification (number of hats and mittens needed)
- Counting by ones and twos
- One pair includes two objects

This activity can be thought of as an introductory experience with functions. In mathematics, a **function** expresses the idea that one quantity (called the input, or argument) can determine another quantity (called the output, or value). A function is sometimes envisioned as a machine that converts the input into the output according to the nature of the function. Each input maps to a specific output. Functions frequently embody a specific rule, but mathematicians emphasize that this is not always the case. In this activity, two mittens are needed for every hat to keep the babies warm. In this situation, the baby is the input and the mittens are the output.

Math Discussions

Because this activity illustrates the interesting pattern of two-for-one, math conversations may revolve around this concept. Teachers should interject into the discussion the relationship of one pair as meaning two objects so that children can think about the idea while modeling with the babies. The conversation should extend to similar situations in the real world. Some examples follow of questions or comments that the teacher might insert into the discussion.

- *Can you help me find one pair of mittens for this baby? Her hands will get cold.*
- *My baby has one pair of mittens; but look, there are two mittens.*
- *How many socks are in a pair?*
- *We have three babies. If each one has a pair of mittens, how many mittens do we need?*
- *Do our babies have more hats or more mittens?*
- *Let's look at what we're wearing. I have one shirt; one pair of pants with two legs; one pair of socks, which is two socks; and one pair of shoes, which is two shoes. What about you?*

ACTIVITY 4.6

Many Ways to Get to Five— A Class Book

Materials

The following materials are needed for this activity:

- Copy of the children's book *12 Ways to Get to 11*, by Eve Merriam (1996)
- White construction paper, 12 × 9 inches
- Assorted collage materials and glue
- White slips of paper, 11¹/₂ × 2 inches, to write the captions

Description

The composition and decomposition of number is an important mathematical concept that children can model in a variety of ways through this project. Although the activity is obviously connected to number sense, the modeling of mathematical situations is also an early algebra concept. Children should be familiar with the Eve Merriam book before embarking on this activity. The book shows many different ways to represent 11, from 6 peanuts and 5 pieces of popcorn to 6 bites, 1 core, 1 stem, and 3 seeds from an apple. Children can use the collage materials to form combinations that equal 5, 6, 10, or any number that the teacher feels is appropriate. After children have completed their picture, they can write or dictate the caption to attach to the bottom of the page. For example, "2 cotton balls, 1 pasta bow, and 2 ribbons make 5." The pages can be bound together to form a class book.

Math Discussions

Children will enjoy comparing the various ways that they have represented a given number. Discussions may ensue when a child misrepresents the number. These conversations should be encouraged, with the teacher in the role of mediator, because they lead children to employ reasoning and proof as a basis for their communication. The following example illustrates how such a conversation might take place.

EXAMPLE 4.7

Lin: "Isaac's picture is wrong because it has more than 5 things."

Isaac: "No, it isn't."

Teacher: "Lin, can you show Isaac how you decided that his picture had more than 5 objects on it?"

Lin points to the collage pieces while counting to 7.

Teacher: "What do you think, Isaac? Are there 7 things on your picture?"

Isaac: "Yeah."

Teacher: "What do you want to do? It's your page. Do you want to just have 5 things in your picture, or do you want to say 'This is a way to get to 7'?"

Isaac: "Get to 7!"

Teacher: "Okay. What should I write?"

(continued)

EXAMPLE 4.7 *(continued)*

> *Isaac: "3 stars, 2 rockets, and 2 balls make 7."*
> *Lin: "And it's 2 more than 5."*
> *Isaac counts the 3 stars and 2 rockets and reaches 5. He looks at the remaining two balls and then nods his head. "Yep, it's 2 more," he says.*

Large-Group Activities

Group-time experiences can be used to highlight small-group math activities before they occur, during their introduction, or after they have been completed, as a follow-up. For collection activities, the teacher might show a group of objects that have already been sorted. Then she can show a new object from the collection and ask the children where it should go, where it cannot go, and why. Group time is also a good forum to discuss alternative ways to sort and classify the materials in a collection.

Patterning activities, such as the "Going Fishing" and "Apples and Pumpkins" activities previously described, also fit well into large-group experiences. The teacher can use the materials to create a pattern and help the children to extend it, perhaps by chanting the pattern along with them. Modeling the activity at group time may increase children's interest in using it at other times of the day, while also providing learning opportunities for children who may not have explored the materials or who need additional support to understand patterning concepts.

Group projects, such as the class book activity "Many Ways to Get to Five," afford children the opportunity to compare many different solutions to the same problem. Sharing a class project reaffirms for children the importance of their work and encourages them to revisit previous experiences. This reflective component helps solidify children's conceptual understanding of the project.

In addition to expanding upon small-group activities, teachers can utilize group time to explore new mathematical situations related to algebra. The activities that follow explore sorting and classifying, patterning, and modeling mathematical situations through whole-group experiences.

ACTIVITY 4.7

Doo Dah Dilemmas

Materials

The following materials are needed for this activity:

- Collection of odds and ends in various sizes, forms, and colors (e.g., key chains, hair accessories, small toys, party favors, sponges, small dishes, lids and bottle caps, refrigerator magnets, junk jewelry)

- Four to six 1-gallon storage bags (clear), mounted to the wall

- Large basket to hold the doo dahs prior to sorting

Today Tehya will sort the "doo-dahs"

Description

Children can sort the doo dah collection throughout the day. During group time, the class can help the teacher understand why various objects are grouped together. In the ensuing discussion, objects may be moved around and regrouped. Each day, children can solve the doo dah dilemma in a different way. New objects, perhaps contributed by children, can be added and other objects removed from the collection. Kindergarten teachers may decide to appoint one child each day to be the doo dah organizer. That child can determine the final placement of the objects prior to the class discussion.

Math Discussions

This activity should generate plenty of conversation as children debate the categories for sorting the objects. It provides an excellent opportunity for children to justify the reasons an object can or cannot go in a particular bag.

ACTIVITY 4.8

Body Patterns

Description

No materials are needed for this activity. The teacher starts by positioning several children to create a pattern, such as forward-backward-backward, forward-backward-backward. With the help of the group, children must decide which way to face when they are called to join the pattern. The activity can be included frequently in group-time experiences. Endless patterning possibilities can be tried, such as standing on the right or the left foot; bending one arm but not the other; standing straight, bowing at the waist, or touching the ground; touching the head, the stomach, or the knee; and so forth. Before the children can extend the pattern, they must discover the attribute being used to create the pattern. As in any interesting mathematical problem, many false starts are possible.

Math Discussions

Conversation will begin with children's suggestions for what constitutes the pattern. This requires them to clearly describe the situation, an important mathematical skill. Many suggestions may be offered before the class finally discovers the pattern. At that point, the discussion must switch to how to extend the pattern. If children give an incorrect response, it is helpful to model the error. As teacher and children chant the pattern from the beginning, they will likely notice and correct their mistake.

ACTIVITY 4.9

Rhythm Patterns

No materials are needed for this activity. Many music educators throughout Europe, Canada, and the United States base part of their instruction on a teaching system devised by composer Carl Orff (1895–1982). The system is based on the idea of **ostinati**, which are repeated rhythmic or melodic patterns. In the beginning, children often copy, extend, and create rhythmic patterns involving clapping, patting the knees, or stamping. Preschool and kindergarten teachers can readily adopt this technique to provide patterning opportunities for their class. Here are some possible patterns:

Clap-pat, clap-pat . . .
Clap-clap-pat-pat, clap-clap-pat-pat . . .
Clap-stamp, clap-stamp . . .
Clap-pat-stamp-pat, clap-pat-stamp-pat . . .
Clap-clap-stamp-stamp-stamp, clap-clap-stamp-stamp-stamp

It is fun, and also challenging, to superimpose the rhythmic patterns over a well-known song. Next, a pattern is integrated into the song "Rain, Rain, Go Away":

Rain,	rain,	go a-	way,
clap	*pat*	*clap*	*pat*

Come a-	gain some	oth- er	day,
clap	*pat*	*clap*	*pat*

We want	to go	out and	play.
clap	*pat*	*clap*	*pat*

Because this activity is mostly about performing the patterns, math discussions are likely to be limited.

ACTIVITY 4.10

12 Ways to Get to 11

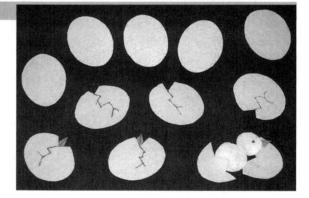

Materials

The following materials are needed for this activity:

- Copy of the children's book *12 Ways to Get to 11,* by Eve Merriam (1996)
- Flannel board
- Cut-outs of the objects used in the book
- Cut-outs of alternative objects (for later use)

Description

This is a perfect book for teachers to illustrate with flannel board pieces. The modeling of the composition of the number 11, used in conjunction with the story, provides many concrete representations and repeated opportunities to count. In subsequent readings of the book, the teacher may deliberately make some errors on the flannel board for the children to catch, which increases their analytical abilities. As a variation, the teacher may provide a variety of flannel board shapes not used in the book for children to use to compose the number 11.

Math Discussions

This activity invites discussion as children analyze what the teacher adds to the flannel board and try to determine the accuracy of the representation. Children may also engage in heated conversations as they decide which (and how many of each) of the new objects can be used to make the number 11.

Algebra Throughout the Curriculum

Opportunities occur throughout the classroom to sort and classify materials; recognize, extend, and create patterns; and model mathematical situations. By supporting the connections between these concepts and children's daily experiences, the teacher helps to solidify foundational algebraic ideas. The examples that follow illustrate important connections between mathematics and play in various areas of the classroom.

Science Area

Symmetry is a special type of pattern that is important in both math and science. It refers to a similarity of form, arrangement, or design, as on either side of a dividing line or around a point. In **reflective symmetry,** if an imaginary line is drawn through an object, one side is the mirror image of the other. A butterfly would be an example. In **rotational symmetry,** an object that is rotated a particular number of degrees around a point looks the same as it did in its starting position. When a square is rotated 90°, it looks the same as it did in its original position. Many plants and animals have symmetrical forms. These materials, or models of them, can be placed in the science area for children to examine and discuss. Insects, leaves, bivalve shells, pansies, crabs, and lobsters are examples of reflective symmetry. Starfish, snowflakes, daisies, and eucalyptus pods are rotationally symmetric.

Music Area

When several sizes of the same instrument are placed in the music area, children often play repeating patterns on them. Teachers can also create patterns and ask children to copy and extend them. The size of the instrument affects the pitch, with larger instruments sounding lower in pitch than smaller instruments. Drums, wood blocks, triangles, and wind chimes can be used for creating patterns. The triangles and wind chimes should be suspended from a frame or pegboard divider so that children can easily move back and forth among the various sizes. Teacher-made instruments can serve the same purpose. For example, hollow metal cans or brass sewing hoops of various sizes can substitute for drums or triangles.

Art Area

Patterns often emerge naturally in children's art work. Children may string beads to form a pattern or draw borders in patterns of colored lines or shapes. An easy way for teachers to introduce the concept of patterns is to supply two sizes of the same tool for children to use in creating paint imprints. For example, the teacher might place two sizes of duck-shaped cookie cutters at the easel. Although some children may automatically begin to create alternating patterns, the teacher might ask others what it would look like if each big duck were followed by a little duck. The question can then change to, "What if each big duck is followed by two (or three) baby ducks? What would that look like?" In this way, different patterns emerge. In future experiences, children may not need the prompts in order to begin creating patterns.

Manipulative Area

Teachers sometimes place small objects or flannel board pieces in the manipulative or book areas of the classroom for children to use in order to re-create stories. When these stories also contain math content, a double purpose is served. Children can model the mathematical situations in the book in addition to modeling the plot. An example would be the well-known Eric Carle (1969) book, *The Very Hungry Caterpillar*. As the story progresses, the caterpillar eats through from one to five specific fruits. Children can use small plastic fruits or flannel-board cut-outs of fruits to recreate the quantities that the caterpillar eats. From a mathematical standpoint, it is better to include more than the requisite number of pieces of each fruit. In this way, children must figure out how many pieces to select to replicate the story.

 ### *Dramatic Play*

In addition to being an excellent area to model mathematical problems that emerge from play (see Chapters 2 and 3), the dramatic play area can be used for sorting and patterning activities. To encourage sorting and classification, the teacher can create labels for food containers in the area. The labels can be changed from day to day. For example, food names might be used one day and colors the next. On Day 1, red, yellow, and green apples might share a basket, whereas tomatoes and apples could be together when color labels are used. Specific materials may be added to the area to encourage patterning. For example, if several colors of barrettes are provided for the dolls, children can use them to create patterns.

UNIVERSAL DESIGN FOR SUPPORTING ALGEBRA

Communication barriers can present problems for some children who are trying to represent mathematical situations. Children with communication challenges may not have the language to express concepts such as more, less, the same, how many, and so forth. In some cases, sign language can help children communicate these ideas. For example, many children know the sign for "more" because it is used in eating situations. This sign can also be used in mathematical situations, especially if modeled by the teacher. In one classroom, line drawings showing the hand position for "more" were used on a spinner for a child with autism spectrum disorder. One "more" drawing appeared on one section of the spinner, and two "more" drawings were placed on the other section. The child used this spinner when playing quantification games, and the images were paired with the actual gesture for the sign. Teachers also might use universal gestures when modeling mathematical situations. Arms wide apart may convey a large amount, whereas hands held close together may represent few. Fingers can be paired with objects when creating signs for specific quantities. It is important to remember that signs and gestures paired with objects and words can be helpful for all children because they create a link to symbolic language.

Another teaching tool that is helpful for children with language delays, as well as for children who are learning English as a second language, is predictable books that convey mathematical concepts. An example is *Blue Sea*, by Robert Kalan (1992). The graphic illustrations by Donald Crews clearly depict the fishes designated as little, big, bigger, and biggest. When teachers point to the illustrations while also reading the brief text, the connection between the words, the illustration, and the concept become clear even to children who do not know the language. One child, who was attending preschool for the first time and knew no English, was able to delightfully shout "ouch" in the appropriate places after only one reading of the book. The book cleverly illustrates a sequence of size relationships, from smallest to largest and back down again.

A key tenet of UDL is the opportunity to represent learning in many different ways. This is also an important concept in algebra; for example, an AB pattern can be illustrated in many different ways. A multisensory approach to representation is helpful for all children and essential for some. For this reason, algebraic concepts such as patterning have been represented through words, rhythm, movements, drawing, and manipulative materials in this chapter.

FOCUSING ON MATHEMATICAL PROCESSES

Although the mathematical process of Representation is a main focus of emergent Algebra activities, all of the process standards are important. The examples that follow highlight some of the ways in which the five processing standards are integrated into Algebra activities.

- *Problem Solving*—When children are presented with patterning activities, they have a problem to solve. What, exactly, is the repeating element? Children may use a variety of strategies to help them recognize and remember the pattern. Examples might be chanting the pattern rhythmically, perhaps with an adult's support, or using small objects to represent the pattern.

- *Reasoning and Proof*—Children working together in groups often receive challenges from their peers. For example, someone may not agree that a particular object belongs in the group in which it was placed. Children may also disagree about whether a mathematical situation has been accurately represented. This was illustrated in the group book project, during which one child felt that a peer had not accurately represented the number 5. When children are challenged in this way, they must rethink what they have done and explain their thinking to others. This requires reasoning and proof.

- *Communication*—When children work together to sort and classify objects, they must agree upon what the categories will be. This requires communication. Similarly, when mathematical representation is incorporated into children's play, as in the "Hats and Mittens" activity, children often discuss the effects. They may notice that for every hat there are two mittens, and discuss this pattern. Some kindergarten children may realize that they can quantify the mittens by counting by twos, but they must count by ones to quantify the hats.

- *Connections*—An important connection that is a goal of algebra activities is for children to use numeric relationships to construct general mathematical concepts. For example, children eventually realize that to add by one, they can simply move forward one unit in the counting sequence; similarly, to subtract by one, they can move backward one unit in the counting sequence. Algebraic concepts thus serve as connectors among the various mathematical content areas.

    Algebra also serves to connect mathematics with other areas of the curriculum. For example, patterning concepts connect mathematics to science, art, and music, and representation of mathematical situations ties math to many play and daily living experiences.

- *Representation*—At the core of algebra is the ability to represent generalized mathematical situations. Young children embark on this concept by representing mathematical concepts in many forms. For example, when they create an alternating pattern with blocks, drawings, and rhythm, they are using three different symbol systems to represent the same concept. This fluency in representation is an important precursor to formal algebraic representation, which will occur much later in their education.

ASSESSING ALGEBRAIC LEARNING

Children demonstrate emergent algebraic concepts of sorting and classifying, patterning, and representation of mathematical situations in many different ways. An easy way to document children's ability to sort by various attributes is to take a digital photograph of their completed activity. A print of the photo can be inserted into each child's portfolio, along with some general notes about the experience. Teachers can also use a simple checklist to document ways in which children sort and classify the objects in a collection.

Patterning activities are often performed with movements, clapping, or rhythm instruments. These must be documented anecdotally. When children create patterns in the art area, however, a work sample can often be saved as documentation. Patterns and representations created with manipulative materials can be preserved with a digital photograph. Finally, when class books are created to show children's representations of a mathematical idea, each child's individual page can be inserted into his or her portfolio when the book is disassembled.

SUMMARY

The Algebra standard encompasses four foundational concepts: 1) understanding patterns, relations, and functions; 2) using algebraic symbols to represent and analyze mathematical situations; 3) using models to represent quantitative relationships; and 4) analyzing change. One of the first relationships that children develop is the understanding of "same" versus "different." They extend this concept when they engage in sorting and classifying activities, such as grouping collections of objects according to a particular attribute. At first, young children may focus on only one attribute of an object, such as color. With experience and teacher support, they begin to realize that objects can be grouped according to a variety of attributes. Eventually, children are able to focus on two attributes simultaneously when grouping objects.

Children who are able to identify specific attributes which are the same or different in objects that are otherwise the same, such as cubes that vary only by color, may begin to organize the objects into pattern relationships. They may decide to alternate red and blue blocks in order to decorate a block structure or create a striped pattern on their artwork. Children may also create and extend patterns through rhythm and movement. An important scaffolding technique for helping children perceive and extend visual patterns is to chant the pattern rhythmically.

Children invent many different ways to represent mathematical situations. They may use fingers or tally marks to keep track of quantities, and they may use objects, drawings, or actions to re-create the mathematical concepts presented in books or discovered through their play. The ability to use a symbol system to represent mathematical ideas is a foundational algebraic concept.

Communication is particularly important in supporting children's development of algebraic concepts. The questions and comments that teachers pose when children are sorting and classifying groups of objects help them develop more flexible thinking. The ability of teachers to connect pattern representations in multiple domains, such as by looking, hearing, chanting, and moving, is critical in supporting children's understanding and development. A representation that children do not understand in one domain may be easily recognizable in another, and the connections help children generalize the concept. Finally, the math-related discussions that teachers have with children throughout the day help them represent and understand mathematical situations and transfer this knowledge to new experiences.

ON YOUR OWN

■ Create a unified collection of objects for children to use as a sorting and classifying activity.

■ Think of four different ways that children could represent the same pattern.

■ Design a gross-motor activity that relates to one of the algebraic concepts discussed in this chapter.

REFERENCES

Baratta-Lornton, M. (1976). *Mathematics their way*. Menlo Park, CA: Addison-Wesley.

Carle, E. (1969). *The very hungry caterpillar*. New York: World Publishing.

Clements, D.H., & Sarama, J. (2007). Early mathematics learning. In F.K. Lester, Jr. (Ed.), *Second handbook of research on mathematics teaching and learning* (pp. 461–555). Reston, VA: National Council of Teachers of Mathematics.

Kalan, R. (1992). *Blue sea*. New York: Greenwillow Books.

Kamii, C. (1982). *Number in preschool and kindergarten: Educational implications of Piaget's theory*. Washington, DC: National Association for the Education of Young Children.

Lester, F.K., Jr. (Ed.). (2007). *Second handbook of research on mathematics teaching and learning*. Reston, VA: National Council of Teachers of Mathematics.

Merriam, E. (1996). *12 Ways to Get to 11*. Fullerton, CA: Aladdin Picture Books.

Moomaw, S. (2008). *Measuring number sense in young children* Unpublished doctoral dissertation, University of Cincinnati. Retrieved from http://etd.ohiolink.edu/view.cgi?acc_num=ucin1204156224.

Moomaw, S., & Hieronymus, B. (1995). *More than counting*. St. Paul, MN: Redleaf Press.

Moomaw, S., & Hieronymus, B. (1999). *Much more than counting*. St. Paul, MN: Redleaf Press.

National Council of Teachers of Mathematics. (2000). *Curriculum and evaluation standards for school mathematics*. Reston, VA: Author.

Pepperberg, I.M. (2008). *Alex and me*. New York: HarperCollins.

U.S. Department of Education. (2008). *The final report of the national mathematics advisory panel*. Washington, DC: Author.

5

Developing Geometry Concepts

The importance of the vertical, the horizontal, and the rectangular is the first experience which the child gathers from building; then follow equilibrium and symmetry. Thus the child ascends from the construction of the simplest wall with or without cement to the more complex and even to the invention of every architectural structure lying within the possibilities of the given material.

Friedrich Froebel (1985, p. 281)

For children, geometry begins with play.

Pierre M. van Hiele (1999, p. 310)

Geometry is another area of mathematics that is considered foundational to later learning. Along with Number and Operations, and Measurement, it is one of the three mathematics content standards designated as a focus for the curriculum in the preschool, kindergarten, and primary grades (NCTM, 2006). Geometry is considered central to the early mathematics curriculum because it is a mathematically important domain, it is an area that young children are developmentally ready to explore, and it connects to the mathematics that they will learn at later grade levels.

Many young children can name simple shapes, and shape-sorting puzzles and manipulative materials are common in preschool classrooms; however, even in the early years, geometry involves much more than simply labeling familiar shapes. As Friedrich Froebel, founder of the kindergarten movement, stressed, through block building, children explore complex concepts such as balance and **symmetry** (Froebel & Jarvis, 1985). The wooden, geometric, building blocks that Froebel designed over 150 years ago are still considered among the best learning materials for preschool and kindergarten children.

THE GEOMETRY STANDARD

Through Geometry, children are introduced to an area of mathematics that is different from, yet related to, Number and Operations, and Algebra. As in early algebra, children must look for similarities and differences among the attributes of objects, albeit now through their explorations of shape and form, and apply patterning concepts in both two and three dimensions. They must also model mathematical situations that involve shape, form, and position and apply their understanding of number to analyzing the properties and locations of shapes.

The Geometry standard articulated by the National Council of Teachers of Mathematics (NCTM, 2000) incorporates four major components:

1. Creating geometric relationships by analyzing the properties of two- and three-dimensional shapes

2. Specifying location and spatial relationships

3. Using transformations and symmetry to analyze mathematical situations

4. Solving mathematical problems through visualization, **spatial reasoning,** and geometric modeling

As do the other content standards, the Geometry standard provides a framework for curriculum and instruction, from preschool through high school and beyond. Preschool and kindergarten are, once again, at the beginning of a grade-level band that culminates in second grade. The expectations for young children are commensurate with their developmental levels and learning trajectories.

Specific goals for young children that relate to the properties of two- and three-dimensional forms include naming, constructing, drawing, comparing, and sorting shapes. In addition, children should begin to describe the attributes of geometric forms and compose (put together) and decompose (take apart) shapes. In the area of spatial relationships, they should begin to describe and name relative positions in space, interpret direction, and specify locations by using terms such as "next to" and "above." Other expectations include recognizing and creating symmetrical shapes, identifying and representing shapes in different perspectives, and recognizing geometric shapes in the environment. The anecdotes that follow illustrate how children develop an understanding of geometric concepts through their play and interactions.

EXAMPLE 5.1

Will and Shunyan had reserved a place in the kindergarten block center, but were instead occupied in the art area. The teacher asked them why they were not using the blocks. "We'll be there soon," they informed him. "We're drawing the plans for our castle." Soon, Will and Shunyan took their papers to the block area and began building. Their completed castle had four identical towers at the corners that were joined by walls of rectangular blocks. A large building that occupied the center of the structure was also symmetrically built.

Children who are prolific builders sometimes make the transition from immediately building with blocks to drawing their structures before they build them. The two boys in this scenario demonstrated strong visualization of geometric shapes, as evidenced by their ability to draw from cognitive memory the shapes and their positioning needed to create the castle they hoped to build. In addition, they were able to represent a geometric structure in both two and three dimensions. Their building showed a feeling for symmetry that young children often represent in their block structures. They begin by building a foundation along the horizontal plane and then add a vertical axis. As they add on to this vertical tower, children often add identical blocks on either side, thereby constructing a symmetrical structure.

EXAMPLE 5.2

Camila watched as a group of children at the special activity table in her preschool class dipped foam geometric solids into paint and pressed them onto paper. "Jada," she asked, "what shape do you think that pyramid will make on your paper?" Jada looked at the side of the shape that she was holding before replying, "A triangle." Jada dipped the base of her pyramid into the paint and pressed it onto her paper. "A square!" she exclaimed in disbelief. "How did that happen?" Camilla encouraged Jada to look at the bottom of her pyramid. "Oh," Jada responded, "the bottom's a square, but the sides are triangles."

One of the expectations of the Geometry standard is that children begin to analyze the properties of two- and three-dimensional shapes. As a result of the design of this activity, along with the carefully timed question of her teacher, Jada begins to analyze the properties of a pyramid, a geometric solid with a four-sided base and triangular sides that meet at a point (see appendix). As a skilled teacher, Camilla has already prepared specific questions to ask when children select particular three-dimensional solids to use in the activity. These questions are meant to guide the children in the analysis of geometric forms.

EXAMPLE 5.3

*Eric watched intently as he pressed geometrically shaped stamps dipped in paint onto one half his paper. When he was satisfied with his picture, Eric folded the blank side over the painted side and carefully rubbed it with his hands. Then he opened the paper. Smiling, Eric showed the other children at his table the shapes that were now on both sides of his paper. "Are they exactly the same?" his teacher asked. "Yes," replied Eric. "Can you show me?" asked his teacher. Eric pointed to a circle and a square on each side of his paper. Then he came to a right triangle. "Wait," he said, "this one's going the other way." "Right," responded his teacher. "That's called a **flip** in geometry. If you take the triangle on the left side of your paper and flip it over, it looks like the triangle on the right side of your paper."*

A flip is one of the transformations that children learn about in Geometry. (More will be said about these transformations later in this chapter.) By designing a fold-over painting activity that involves shapes, the teacher helps children visualize what a transformation looks like. This type of activity also creates a symmetrical image, another important geometric concept.

EXAMPLE 5.4

Matt could not find the new truck that he had brought to preschool. "Did you look in your cubby?" his teacher asked. "Yes," Matt replied, brushing away a tear. "I'll help you look," the teacher said. Soon the teacher spied the missing truck. "Matt," he said, "I see it. It's in the cubby next to yours, on the top shelf above Susie's shoebox."

Opportunities to help children learn positional terminology, such as "next to," "on top of," and "above" in the preceding example, occur throughout the day. Hearing these words within a meaningful context, as in the hunt for a lost toy, reinforces their meaning for children. Understanding location words is another important geometric concept. Teachers should remember to use them often.

THE DEVELOPMENT OF GEOMETRIC REASONING

A considerable amount of research exists regarding children's understanding and development of geometric concepts. This research encompasses children's understanding of spatial relationships, including their ability to navigate through their environment and locate objects, as well as their ability to visualize and mentally manipulate two- and three-dimensional forms. Considerable research has also been conducted related to children's perception and knowledge of shape (Clements & Sarama, 2007). In addition, educator Pierre van Hiele has developed levels of thinking that inform teachers of expected trajectories in geometric understanding (van Hiele, 1999).

Spatial Reasoning

Spatial reasoning refers to the ability to visualize and mentally manipulate spatial patterns. Individuals use this ability in daily life to mentally plan which route to take to work or school, perhaps comparing several possibilities; to organize shelves and cupboards to contain an array of items of various shapes and sizes; to load the dishwasher; to envision a rearrangement of furniture in a particular room; to work a 1,000-piece puzzle; and so forth. Spatial reasoning is directly involved in many professions, including architecture, engineering, art, science, and mathematics, and development of spatial reasoning is related to general achievement in mathematics (Wheatley, 1990).

Two important components of spatial reasoning are **spatial orientation** and **spatial visualization** (Clements & Sarama, 2007). Development of these concepts begins early in life and continues for a prolonged period.

Spatial Orientation

Spatial orientation refers to an understanding of the body's position in relationship to physical space and the ability to navigate through that environment. According to Piaget, infants are born without an understanding of physical space or of object permanence; however, through active movement and manipulation of their environment, they begin to develop these concepts, first with regard to space that is near them, and later space that is farther away (Piaget & Inhelder, 1967, 1969). This contention is supported by more recent research (Haith & Benson, 1998).

As children begin to navigate through space, they develop a pattern of movements associated with a particular goal. For example, an infant may pull herself up in her crib to stand facing the door that Daddy always comes through early in the morning. When crawling or walking, older infants and toddlers update their location according to familiar landmarks, such as knowing to move around the sofa to get to the dog dish. One 10-month-old was observed making the daily early-morning rounds of trash dumping. He regularly crawled to waste baskets in the bedroom, living room, and kitchen so that he could upend the baskets and explore their contents, an activity which had to be carefully monitored by his parents. This development of spatial reasoning continues over time, with preschool children able to plan organized searches for objects within their immediate environment (DeLoache, 1987).

Related to spatial orientation is the use of symbol systems, including language and visual representations, to denote location. Of particular interest to preschool and kindergarten teachers is information on the acquisition of spatial terms, such as "in," "on," "beside," and "in front of," because the understanding of these terms is specifically cited in the Geometry content standard. The order in which children learn these terms is consistent across languages (Bowerman, 1996). "In," "on," and "under," terms that relate to an object in direct contact with another, are the first spatial terms acquired, along with the movement terms "up" and "down." Because babies are frequently picked up or put down, it is perhaps not surprising that these terms are learned so early. Next, children learn words of proximity, such as "next to," "between," and "beside." These terms are followed by the acquisition of words related to position in reference to another object, but not necessarily in close proximity, such as "in back of," "in front of," and "behind." As both parents and teachers can attest, directional terms such as "left" and "right" are learned much later, often not until first, second, or even third grade (Clements & Sarama, 2007).

Young children also begin to use and interpret visual symbols to represent spatial relationships within their environment. For example, preschool children use blocks to reconstruct familiar structures such as houses. Within these buildings, they place dollhouse toys in configurations that resemble their real-life experiences, with appliances in the kitchen and trees and cars outside. In research, after having been shown the hiding spot of a toy dog on a model, a photograph, or a line drawing of a room, 3-year-old children were able to find a toy dog hidden in the same location in an actual room (DeLoache, 1987). By age 4, many children can interpret some of the symbols used on a map, such as an X on a treasure map, but even by age 5 or 6 they may not understand how the locations on the map correspond to locations in the real world that the map represents (Liben & Yekel, 1996).

Spatial Visualization

Spatial visualization is the ability to mentally produce and manipulate visual images of two- and three-dimensional objects. At first, these images are **static images.** This means that, although children can recall the image and even mentally examine it, they cannot mentally move the image. For example, when many 3-year-olds hear the word "cat," they can visualize and even describe a cat, but cannot mentally move it to determine whether it would fit into a shoebox. The ability to conceptualize a **dynamic image,** which is an image that the individual can mentally move, develops later and incorporates the common geometric transformations of **turn,** flip, and **slide.** A turn involves rotating an object, such as when a child turns a puzzle piece to try to fit it into a designated space. A flip, which is also called a **reflection,** is the mirror image of an object. If a child uses a page-turning motion to move an object, the flip side will be visible. Finally, a slide, or **translation,** indicates that an object has been moved without having been flipped or rotated so that every point on the object moves the same distance and direction. Sliding a table along the wall would be an example.

Piaget maintained that most children do not have full dynamic images until about age 7; in other words, they cannot mentally envision all of the transformations previously described until that age (Piaget & Inhelder, 1967). Research conducted with second-graders supports Piaget's argument; the students were able to learn manual procedures for performing transformations, but were unable to apply them to mental transformations (Williford, 1972). Notwithstanding, more recent research indicates that young children may internalize some images of dynamic motions (Clements, Battista, Sarama, & Swaminathan, 1997).

There appears to be a hierarchical order in children's abilities to represent geometric images. First, children learn to create a matching shape when the model is clearly visible. Next, they are able to do this from memory, with the model not present. This is followed by representations that reflect **rotation** or by the taking of another's visual perspective. Preschool children are able to create representations only at the first two levels (Rosser, Lane, & Mazzeo, 1988). By the end of kindergarten, however, 52% of the children in a study in Australia were able to visualize the effects of motions (Clements & Sarama, 2007). This shows that young children are able to make substantial improvement in this area, particularly if teachers support this learning.

Geometric Shapes

Preschool and kindergarten children are highly successful at identifying common shapes. In a study conducted with a large sample of children 4, 5, and 6 years of age, 92% of 4-year-olds, 96% of 5-year-olds, and 99% of 6-year-olds identified a line drawing of a circle (Clements, Swaminathan, Hannibal, & Sarama, 1999). Most children, however, were not able to describe the circle as anything other than "round." In the same study, 82% of 4-year-olds, 86% of 5-year-olds, and 91% of 6-year-olds could identify a square. The children were less accurate when identifying triangles (60%) and rectangles, although these results were not remarkably smaller than those of elementary students (Clements, 2004). This raises a concern that, although children have strong initial competencies, traditional instruction does not seem to extend their knowledge. One problem is that children usually are presented shapes in only a prototypical fashion. For example, triangles are usually **equilateral** or **isosceles,** with a horizontal base. This leads children to assume that these are the only forms that can be labeled "triangles."

Dutch educator Pierre van Hiele proposed a series of thinking levels that children progress through as they develop their understanding of geometric forms. His view is that this progression is critical for teachers to understand; otherwise, their instruction may create a gap in students' understanding, which they cannot bridge. According to van Hiele, instruction should foster development from one level of thinking to the next through a sequence of activities that begins with exploration. Gradually, concepts and related language should be added. Finally, summary activities should be introduced that help students integrate into their existing knowledge the new concepts they have learned (van Hiele, 1999). The following section covers the levels of geometric thinking outlined by van Hiele (1986, 1999).

van Hiele Levels of Geometric Thinking

1. *Visual Level*—At this level of thinking, children judge a figure according to its appearance. Children may say that a shape is a square because it "looks like a box," or label it a triangle because it is "pointy." Children often have a prototype in mind and reject any shape that does not fit this example. For this reason, a square that is rotated 45 degrees may be called a "diamond" and not a square. Conversely, a triangle with slightly rounded sides may be deemed sufficiently like a prototypical equilateral triangle to be labeled as a triangle.

2. *Descriptive Level*—Children at the descriptive, or analytic, level begin to use language to describe the properties of shapes. For example, they may reason that a figure is a triangle because it has three straight sides all connected by points. On the basis of these properties, children develop mental classes of figures. At this level, however, children are not yet able to place these properties in a logical order to create precise mathematical definitions. Therefore, they may still insist that a square is not a rectangle.

3. *Informal Deduction Level*—It is not until the informal deduction level that students are able to order properties logically. According to their understanding of the properties that they know, they formulate definitions and use them to judge the relationships among figures. It is at this stage that students understand why a square is also a rectangle.

4. *Rigor*—At the final level of reasoning, students are ready to understand the formal deductions that are essential for learning Euclidean geometry. In van Hiele's opinion, students frequently have difficulty in formal geometry courses because they have not yet mastered informal deductive reasoning.

In preschool and kindergarten, children are usually at the first level of van Hiele's hierarchy. The goal for teachers, therefore, is to provide experiences that support visualization concepts and help children transition from the visual to the descriptive level.

Clements has proposed some elaborations to the van Hiele model (Clements & Sarama, 2007). He suggests a level before van Hiele's visual level to describe children who cannot yet reliably distinguish particular shapes, but are starting to form visual schemes for the shapes. Clements also argues that van Hiele's visual level includes both visual images and declarative statements; therefore, he does not view the levels as strictly visual progressing to verbal.

DESIGNING THE GEOMETRY CURRICULUM

Based on what is known about children's development of geometric thinking, as well as recommendations from the NCTM Geometry standard, the geometry curriculum in preschool and kindergarten should center around the following areas:

- Recognizing and analyzing the properties of two- and three-dimensional forms, including their functions in the real world

- Understanding symmetry and the effects of transformations (flips, turns, and slides)

- developing spatial reasoning and specifying location

- Solving mathematical problems through spatial reasoning and geometric modeling

In order to meet these goals, children need carefully planned hands-on experiences with geometric materials, accompanied by related vocabulary and interesting mathematical discussions. As with number and algebra concepts, math talk related to shapes and the relative positions of objects should be interjected regularly into teachers' conversations with children. Small- and large-group experiences draw needed focus to geometry in preschool and kindergarten programs, and incorporation of geometry throughout the curriculum helps children connect geometry to other areas of interest.

Math Talk

Two types of math talk are related to geometry: 1) discussions involving geometric shapes, and 2) conversations related to location and spatial reasoning. Although it is certainly appropriate for young children to use common terminology for geometric forms, such as *ball* for a sphere or *box* for a square, teachers should know and model the correct mathematical terms. For this reason, illustrations of two- and three-dimensional figures, along with their correct labels, are included in the appendix. These terms should become part of the regular vocabulary of teachers, much as labels for zoo animals, types of trucks, or dinosaurs are for many preschool and kindergarten teachers. This takes practice. For easy access, teachers might want to copy onto note cards the drawings and terms that are more difficult for them to remember. If teachers are not comfortable with their own knowledge of geometric forms, they are likely to shy away from talking about geometry with children.

Teachers regularly mislabel some shapes. One common example is a square or a rhombus that is rotated 45°. Both children and teachers often call this shape a diamond, which is not a mathematical term. Without correcting the child, teachers should simply point out that in math, that shape is called a square (or rhombus), and one can tell by turning it. Other shapes that are frequently mislabeled are the **ellipse,** which is often confused with an **oval,** and the **kite,** which is frequently mislabeled as a diamond. As in all areas of their education, it is helpful for children to learn the correct terminology from the beginning rather than to have to relearn it later.

Another geometry topic that teachers rarely talk about in the early grades is lines. This is unfortunate because the world is full of lines, and they are an important property of shapes that children must begin to distinguish as they advance from the visual level of geometric reasoning to the descriptive level. Teachers can talk about lines that are straight versus curved and can help children identify and sort shapes on the basis of that distinction. Another linear relationship involves whether lines intersect or are parallel. Lines on the same plane that never cross are parallel; floor boards, the edges of shelves, and railroad tracks (both the rails and the ties) are examples. Intersecting lines meet at some point; the corners of a table, scissor blades, and fence boards meeting fence posts are examples likely to be familiar to young children. Because children regularly use lines in their drawing and writing, these activities provide teachers with many opportunities to comment on line characteristics: straight, curved, parallel, intersecting, and so forth.

A final geometric topic that is important to include in conversations with young children is spatial relationships, including specific location. Positional terms should be regularly incorporated into discussions with children. Perspective can also be included. For example, a car that is seen as behind a block from one position may be seen as next to it or even in front of it from other locations. A conversation might go something like this: "Brian, from where I am sitting, I can see a cow in front of your barn. Where is the cow from where you're sitting? Can you even see it?"

All of these conversations help to bring geometry into the everyday experiences of young children and help children to learn the related vocabulary, within meaningful contexts. Math talk related to specific curriculum is highlighted under each activity in the sections that follow; however, conversations related to geometry should not be limited to only a few specific situations. Teachers should remain alert for opportunities to interject geometry-related topics into the play and daily classroom experiences of children in order to maximize their opportunities to learn these foundational concepts.

Individual or Small-Group Activities

There are many interesting manipulative materials available that are related to geometry. These materials are often used by individual children to explore shape relationships, but children can also use some of them in pairs or small groups. Geometric manipulative materials should be consistently available in preschool and kindergarten classrooms. Table 5.1

Table 5.1.   Geometric materials and related concepts

Material	Description	Geometry concepts
Shape sorters and puzzles	Children manipulate shapes (turning, flipping, and sliding) to fit them into corresponding spaces.	Shape recognition Transformations
Pattern blocks	Children combine shapes to recreate pictures or patterns or to create their own designs.	Shape recognition Shape composition Geometric patterning
Attribute blocks	Children can sort the blocks by shape, size, and thickness.	Sorting and classifying Shape recognition
Geometry building sets	Children use either flat shapes or sticks and connectors to construct geometric solids.	Composition and decomposition of three dimension forms Shape recognition (two and three dimensions)
Geometry nets	Children can compose and decompose geometric solids by folding and unfolding the patterns.	Composition and decomposition of three dimension forms Shape recognition (two and three dimensions)
Geo-boards	Children stretch rubber bands between prongs on a board to create shapes, parallel lines, and intersections.	Shape composition Line segment composition

describes some of the more common geometric manipulative materials and the concepts that they are designed to support.

In addition to making traditional manipulative materials regularly available, pre-school and kindergarten teachers should plan small-group activities that focus on specific geometry concepts. Small groups enable teachers to scaffold learning for individual children, collect assessment data, and lead math discussions. The information obtained from small-group interactions is critical for teachers in planning the ongoing geometry curriculum.

As previously indicated, a well-stocked block area is a strong contributor to geometry development. In fact, most of the geometry curriculum could be taught in the block area. Children learn to recognize shapes and, through their building activities, explore the properties of geometric solids. They also develop concepts of balance, symmetry, pattern, and structural support. In addition to building with the blocks, children often put shapes together to compose new shapes. For example, two triangular building blocks can be combined to form a rectangle, and two squares form a rectangle that is no longer a square. The language that accompanies block building can focus on positional terms because children often place block accessories, such as animals, furniture, or people, at various places within their block structures.

A final component of the Geometry standard is to use visualization, representation, or modeling to solve mathematical problems. This concept is embedded in many geometry activities. For example, with geometric building materials such as Polydrons, the mathematical problem is how to combine two-dimensional shapes to create a three-dimensional shape. With shape sorters and puzzles, the problem is how to fit each shape into a designated space. In the specific activities described next, the mathematical problem involved is emphasized.

Two- and Three-Dimensional Shapes

The geometry curriculum in preschool and kindergarten is often limited to simply recognizing and labeling two-dimensional, and maybe three-dimensional, shapes. Such a limited presentation of geometry constrains children in their development of concepts such as the components of geometric figures, the relationship among shapes, and the importance of geometry in solving real-world mathematical problems. Even worse, it is boring. Teachers often express frustration about what to add to the geometry curriculum to make it more interesting for children. One teacher complained, "When I put out the geometric solids, the children pick them up once and look at them, and then they're done."

Children are interested in materials that they can manipulate to cause a noticeable change. One way to add interest to shape materials is to move them to the block area. Children may use flat shapes to create floors for their buildings and in this way discover the various ways that shapes can fit together. They may compare the conventional geometric solids to similar shapes among the unit blocks. This invites discussion about similarities and differences among three-dimensional forms. Children can even experiment with the physical properties of geometric solids by creating a ramp with the blocks and observing how three-dimensional objects react when placed on an incline. Some of the objects are particularly interesting; for example, the cylinder rolls when placed on its curved side, but not when set on a flat end. The cone is fun to watch because it rolls off to the side. Shape materials can also be placed in the art area. Children may trace around flat shapes to create geometric patterns or dip each **face** (side) of geometric solids into paint to see what shapes they create. All of these activities move children toward the descriptive stage of geometric thinking. A simple change of venue for geometric shapes can immediately upgrade a stagnant curriculum.

The activities that follow focus on the composition and decomposition of shapes and the relationship among geometric forms. They serve as models for a wide range of activities that teachers can develop to expand their geometry curriculum.

ACTIVITY 5.1

The Magnificent Shape Quilt

Materials

- The following materials are needed for this activity:
- Felt squares, 6 × 6 inches, cut from felt in bright colors
- Small geometric shapes, in standard and nonstandard configurations, cut from bright colors of felt with a self-adhering backing
- Puffy paint, to use to write each child's name on the finished quilt squares

Description

Each child can experiment with positioning shapes to create an individual collage. Older preschool and kindergarten children should be encouraged to arrange and rearrange their shapes until they are satisfied with the result. At this point, the backing can be removed and the shapes mounted to the felt squares. (Younger children may be more interested in the process of mounting the shapes to the backing than in creating a specific piece of artwork.) As children engage in the process of creating their quilt squares, they may observe the ways in which shapes fit together to create larger shapes. Children may also create symmetrical or linear patterns, form representational images, or simply design a random shape collage. From a mathematics perspective, the conversation that accompanies the activity is a critical component; therefore, teachers should plan for an adult to be seated with the children as they work on their designs. Once the squares are completed, each child's name can be added with puffy paint. The quilt squares can be whip-stitched together or mounted to a backing to create a class quilt.

Quilts provide a fascinating array of geometric representations for children to study. It is helpful if children have the opportunity to examine quilt designs, either on actual quilts or in photographs, before they work on this project. Calendars, books, and Internet web sites are good sources of photographs of quilts. Many children's books have a quilt theme and provide an excellent connection to this activity. A list is provided in the integrated curriculum section of this chapter.

Because this activity uses geometry as a creative endeavor, there is no specific math question to solve; however, mathematical problems may arise from teacher comments or questions. For instance, during the exploration process, the teacher might ask children what they could make with four triangles, two rectangles, a circle and a triangle, and so forth.

Math Discussions

Many excellent topics of conversation should emerge during this project. Children who are just learning the names of shapes should have this vocabulary reinforced by the teacher. During the activity, children should be encouraged to experiment with placing some shapes so that they touch other shapes, although the final design should, of course, be the child's choice. When children combine shapes, this should be discussed. The teacher might say, "What happened when you put a square and a triangle together?" The child may reply that she created a house, and the teacher can agree while adding that in geometry that shape is called a "pentagon." Teachers should also look for symmetrical relationships that children may create. Some children may begin with a central shape and then branch out, in a symmetrical fashion, from that starting point. Other topics of conversation might be items that are the same shape, but a different size; how shapes look when they are turned (rotated), or flipped; and the types of lines that form the edges of the shapes.

ACTIVITY 5.2

Quilt Bingo

Materials

The following materials are needed for this activity:

- Quilt-square game board for each child, created by tracing quilt designs that feature geometric shapes onto paper and mounting them to poster or tag-board
- Shapes that match all those used on the game boards, made by cutting the shapes from heavy-weight, colored paper and laminating (scrapbook paper is a good source)
- Grab bag, to hold the shapes

Description

This activity coordinates with the previous quilt activity and may serve as a precursor to it. Children take turns passing the grab bag and drawing shapes to match those on their game boards. The matching process encourages children to focus on the attributes of the shapes, especially when they are searching for a particular shape to complete their boards. In the process of playing the game, children may observe how particular shapes are combined to create interesting patterns.

Math Discussions

Once again, the discussions that occur during this activity are a critical component of the learning. Without the accompanying math talk, children may match shapes, but not learn their names or notice the patterns that they create. A main focus of the conversation should be on the attributes of the various shapes. For example, if a child is looking for a triangle, the teacher might ask whether it will have straight or curved sides, and how many points it will have. After several teacher-led clues, children may begin on their own to look for particular attributes.

In this activity, the mathematical problem is how to use the attributes of shapes to find particular shapes in the grab bag and then match them to the board. In many cases, children will need to use transformations to fit shapes into their proper spaces. Teacher scaffolding may be necessary, especially when flips are involved.

This activity also provides an excellent opportunity to talk about transformations. As children draw shapes from the grab bag, at first they may not be able to visualize where the shape can fit on their game boards. The teacher may need to scaffold, such as suggesting that the child turn the shape and look at it again. Other children can be brought into the conversation. For example, the teacher might say, "Jay, can you see a place where Margaret's shape will fit on her board? What does she have to do to make it fit?" Conversations such as this help children not only to visualize the results of transformations but also to learn the accompanying terminology.

Transformations

In their interest to help children learn to name shapes, teachers may forget that there are other important geometric concepts to emphasize on a regular basis. The concept of the **transformation** is one of them. The area of the curriculum in which children are most likely to use transformations is the manipulative area, especially when they work puzzles. When helping frustrated children fit puzzle pieces into their appropriate places, teachers may, in fact, suggest transformations without realizing that these are geometric concepts. As examples, teachers may say, "Turn your piece around," "Slide it over here," or "Flip it over and try it again." These statements all illustrate uses of geometric transformations.

Teachers can extend children's understanding of transformations by drawing attention to them in other situations. For example, children are sometimes given cookie cutters in various shapes to use with play dough. Teachers can use this opportunity to investigate transformations with the children. The teacher might ask, "If you turn your triangle a little bit, what does it look like? Is it still a triangle?" Leading questions such as these help children realize that transformations do not change the essential components of shapes. Through their explorations, children become more flexible at recognizing shapes in different orientations. The activity that follows is one example of the types of experiences that teachers can plan in order to focus on transformations.

ACTIVITY 5.3

Shape Aliens in Action

Materials

The following materials are needed for this activity:

- White construction paper, 9×12 inches
- Shapes, cut from colored construction paper
- Glue

Description

This activity begins with children creating their own imaginary "shape alien" by gluing the colored shapes onto white construction paper. Because these aliens are made of shapes, they can move by flipping, turning, or sliding (geometric transformations). On subsequent days, children can use their shape aliens to model these transformations, which is the mathematical problem for this activity. The four alien orientations can then be displayed together.

In order to flip their alien, children must place a matching shape on top of each shape that constitutes the alien. Once this is done, children put either a dot of glue or a double-sided photo tab on each shape. Next, a blank sheet of white construction paper is placed directly on top of the artwork and carefully rubbed so that the shapes underneath will stick to it. When the top paper is turned over, the shape alien has flipped! Minor adjustments can be made if some shapes have shifted position.

To turn and slide the shape alien, two color copies of the original artwork are needed. These can be created with a photocopy machine, or scanned into a computer and printed.

Children can turn the alien relatively easily by simply rotating the paper. To slide the alien, they need to keep it in its original, upright position, but place it in a different spot on a new piece of paper. A geometric translation (slide) requires that all points on the shape move the same distance and direction. If the shape is turned or flipped, some points will move farther than and in a different direction from other points. To make their alien slide, children can cut around the outside of the alien and glue it to the new paper, perhaps farther to the left or right, or nearer the top or bottom of the paper. The four versions of the shape alien can be appropriately labeled and taped together for display. They can also be stapled together with a front and back cover to form a shape alien book.

Math Discussions

The conversation surrounding this activity, of course, focuses on transformations. Teachers should encourage children to describe how their alien moves as it flips, turns, and slides. Specifically, children should observe the positions of the various shapes that the alien comprises. To encourage these observations, teachers should ask questions such as, "What happened to this purple rectangle when your alien turned?" Displaying the aliens encourages children to continue discussing transformations as they compare aliens among their peers.

Spatial Relations

Recall that spatial reasoning includes an understanding of the individual's position in space, the ability to navigate through that space, and the specific location of other landmarks and objects relative to the individual's position. A further component of spatial reasoning is the ability to visualize and mentally manipulate objects. Visualization is a strong component of activities such as the shape aliens project just described. Therefore, this section focuses on spatial orientation.

Teachers have many opportunities throughout the day to focus attention on spatial orientation and navigation through space. As children play, teachers can use positional words to describe their actions. For example, on the playground, the teacher might comment, "I saw you go over the top of the balance beam and behind the climbing cube to find your ball." As indicated earlier, teachers should regularly include positional statements in their conversations with children. An example would be, "Can you bring me the stapler? It's on the middle shelf next to the tape." The activity that follows focuses on spatial orientation in two ways, first through children's actions, and then through their recollections as they view photographs of their actions. (First get permission from parents to photograph students.)

ACTIVITY 5.4

Where Are Our Friends?

Materials

The following materials are needed for this activity:

- Note cards, 4 × 6 inches, with directions printed on them that tell children where to go for their photograph

- Digital camera

- Manila or construction paper, for mounting the photographs and printing descriptions of where the child is located (e.g., Brittany is sitting under the ladder climber)

Description

This activity can take place anywhere in the classroom, but the gross-motor room or playground are ideal because children can be positioned on top of, under, behind, and next to the play equipment. For the first part of the activity, children draw a note card with instructions for where and how they should position themselves for their photograph. Here are some suggestions:

• Stand on top of the climbing cube.

• Lie down under the balance beam.

• Sit under the picnic table.

• Kneel behind a tricycle.

The teacher can help the children read their cards and evaluate how well they interpret the directions. Support should be given as needed. Once a child is positioned as directed, the teacher can take a digital photograph. These can be printed and mounted on paper.

The second part of the activity involves interpretation of the photographs, which again requires the use of positional terms. Children are shown their photographs and asked to describe exactly where they are in the picture. Teachers can ask questions or make comments that focus on positional terms. The descriptions can be printed, by either the teacher or the child, on the appropriate page. All of the pages can then be compiled into a *Where Are Our Friends* book.

In this activity, the first mathematical problem is for children to find the exact location indicated on their direction card. The second problem is to describe that position in the photograph.

Math Discussions

This activity is designed to promote discussions about location and spatial orientation. The directions on the note cards may initiate the conversations. Children may have questions about what the location given on the card means, and teachers can break down the directions into parts. For example, if the card asks the child to sit behind the trampoline, the teacher might first ask the child to find the trampoline, and then discuss what "behind" means. Further conversations will ensue when children look at their photographs. At this time, teachers can ask additional questions related to location and orientation. For example, if a child is standing next to the climber, the teacher might ask where the child would be if she moved to another position or what she would look like if she turned her body around. Children are likely to read the class book to one another for many days. Each reading induces more conversations.

Large-Group Activities

Small-group activities provide an excellent forum, and perhaps the most ideal one, for moving children forward in their understanding of geometry. Because there are so many curriculum demands, however, geometry is only one of many topics that teachers must plan for small-group activities. Nevertheless, by devoting a portion of group times to geometry, teachers can solidify developing concepts. The activities that follow can be incorporated daily for an extended period or used on an occasional basis.

Two- and Three-Dimensional Shapes

In keeping with the goals articulated by NCTM in their Geometry standard, large-group activities related to geometric shapes should focus on their properties and how the shapes appear in various orientations. In order for children to develop categories related to shapes, it is important for teachers to represent shapes in a variety of configurations as well as orientations; for example, triangles that have sides and angles of various sizes should be included, and the base should not always be at the bottom. The activities in this section can help children to recognize shapes, regardless of configuration and orientation, and to discuss their properties.

ACTIVITY 5.5

Weekly Geometry

Materials

The following materials are needed for this activity:

- Magnetic shapes, inexpensive and commercially available
- Magnetic board
- Large geometric shapes, in a variety of sizes and configurations, cut from poster board
- Shape cards, made from note cards, each with a picture of a shape and its name printed on the card
- Three bags, to hold the two types of shapes and the shape cards

Description

Three geometry activities that can be used on a weekly, or even daily, basis in group times are included here: *Grab Bag*; *Flip It, Turn It, Slide It*; and *Shape Sort*. For the Grab Bag activity, several children are selected to draw a shape card, which they share with the class. The children then try to find their shape in the grab bag without peeking. After they remove a shape from the grab bag, it can be compared with the card.

For the Flip It, Turn It, Slide It activity, several children are selected to pull one of the large poster-board shapes from the bag. After identifying their shape, with help from the class if needed, the children bring their shapes to the front of the group and demonstrate how the shape looks when they flip it, turn it, or slide it.

For the Shape Sort activity, the magnetic shapes are placed randomly on a magnetic board. Teachers should place the shapes so that they are not all in a standard orientation; for example, rectangles may be positioned so that the longest sides are horizontal, vertical, or on a slant. Children then offer suggestions for ways to sort the shapes. After one of the suggestions is adopted, children help the teacher decide which shapes belong together. Some possible ways to sort the shapes are curved versus straight sides, all shapes of the same type grouped together, and triangles versus not triangles.

Math Discussions

Even though not every child has a chance to pick a shape each day, teachers should involve all of the children in the related math discussions. For Grab Bag, the teacher might ask the class to provide suggestions for the child who is feeling inside the bag for a particular shape. For example, if a child is searching for an ellipse, the other children might suggest that the child try to find a shape with curved edges; for a rectangle, they might suggest locating a shape with straight sides. The teacher might prompt with a question regarding the number of straight sides that the rectangle must have. If a child who is looking for a rectangle pulls out a square, the children may argue that it is not a rectangle. This gives the teacher the opportunity to explain that a square is a special kind of rectangle, but it is still a rectangle.

Teachers should directly involve children in the shape-sorting activity, through a class discussion. As the teacher points to each shape, children should be asked why it fits or does not fit into a particular category. The teacher can also guide, and perhaps list, the children's ideas about sorting categories. Each day a different suggestion can be implemented.

ACTIVITY 5.6

Class Shape Big Book

Materials

The following materials are needed for this activity:

- Digital camera
- Construction paper, 12 × 18 inches
- Primary lined paper, to write descriptions of the pictures

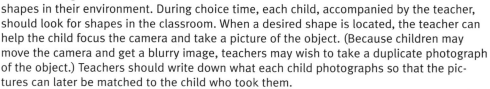

The doors to the playground shed had rectangles at the top and octagons inside of rectangles on the bottom. There were triangles in the corners.

Julie and Orion

Description

This activity is designed to help children recognize and remember shapes in their environment. During choice time, each child, accompanied by the teacher, should look for shapes in the classroom. When a desired shape is located, the teacher can help the child focus the camera and take a picture of the object. (Because children may move the camera and get a blurry image, teachers may wish to take a duplicate photograph of the object.) Teachers should write down what each child photographs so that the pictures can later be matched to the child who took them.

Once everyone has had a turn to take a shape photograph, the pictures can be printed and mounted to the top half of the construction paper. Lined paper can be attached to the bottom portion of the construction paper, for children who prefer to write on lines. Children can discuss their photograph with the teacher and decide what to write on their page of the shape book. The pages can then be assembled into a class big book to share at group time.

Math Discussions

Children will probably be eager to discuss their pages in the *Class Shape Big Book*. Children like to read big books more than once; so, in future readings, the teacher can generate additional conversation about the photographs. For example, the teacher might ask whether other shapes that are not mentioned in the text appear in a particular picture. Some examples follow of other questions that might be introduced to provoke discussion.

- *This page says that the clock is round. What other shapes can you think of that are round?*
- *Kim's page says that the seat of the chair is a trapezoid. Do you think that all of the chairs in our class have trapezoidal seats?*
- *I've been wondering what shape appears the most in our book. How could we find out?*

Transformations

Transformations have already been highlighted in several activities, including Quilt Bingo, Shape Aliens, and Daily Geometry. The activity that follows provides yet another way to focus on shape identification, matching shapes, and transformations, as children work together to dress Mr. and Mrs. Teddy Bear.

ACTIVITY 5.7

Dressing Mr. and Mrs. Teddy Bear

Materials

The following materials are needed for this activity:

- Two large teddy bears, traced onto light brown flannel or felt
- Felt shapes, to match the shapes on the outline, for dressing the bears
- Basket, to hold the shapes

Description

Dressing Mr. and Mrs. Teddy Bear is a delightful way for children to focus on shape. During group time, the basket is passed around, and each child selects a shape. Children take turns adding their shapes to the bear outlines, which are mounted to the wall or bulletin board. Teachers can facilitate this process in one of two ways: 1) Children can bring their shapes to the board and find a space where they fit, or 2) the teacher can point to a shape on the bear outlines and ask who has a shape that may fit there. As children add their shapes to the bears, opportunities to label shapes in various configurations abound. For example, Mr. Teddy Bear has a feather in his hat that is a long, skinny triangle—not the familiar equilateral or isosceles triangle that children usually see. In order to match their felt shapes to the drawing, children will need to perform transformations. For example, Mrs. Teddy Bear has a flower made of ellipses in her hat. Children must rotate these ellipses to make them fit the flower. In other cases, children may need to flip a shape. Mr. Teddy Bear has pant legs that are trapezoids. Because the trapezoids are not symmetrical (isosceles), the child with the second pant leg may need to flip it in order to make the trapezoid fit.

Math Discussions

This activity generates considerable discussion, as children are eager to talk about the shapes and to coach one another as they try to dress the bears. In addition to providing the necessary vocabulary for shapes that children do not recognize, such as trapezoids and parallelograms, teachers can help children focus on transformations. The activity can stay on the board for children to return to during choice times.

Spatial Relations

Yet another area of geometry that can be included in group-time experiences is spatial relations. Teachers might take a toy and place it in various positions relative to another object, such as a truck. Children become much more interested when these movements are added to a song. Using a toy pig, a truck, and the familiar tune of "The Farmer in the Dell," the teacher might sing as follows:

> *The pig is in the truck,*
> *The pig is in the truck,*
> *Hi, ho, the derry-o,*
> *The pig is in the truck.*

Other positional terms can then be inserted into the song. For example, the pig might be *in front of*, *next to*, *in back of*, and *under* the truck. Children readily join in the singing of songs such as this and seem to focus longer on musical activities than on spoken ones.

Later, in elementary school, children will learn to locate positions on a graph. As a precursor, the next activity positions objects on a 3 × 3 matrix. Children can use clues or maps to locate the objects.

ACTIVITY 5.8

Pet Hospital

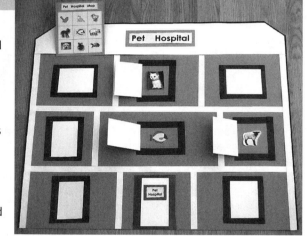

Materials

The following materials are needed for this activity:

- Pet hospital (as pictured), made from poster board that is divided into nine sections
- Construction paper, to create doors for the rooms in the pet hospital
- Illustrations of a puppy, cat, bird, rabbit, turtle, lamb, goat, and fish
- Velcro or magnetic tape, to attach the illustrations to the poster board
- Maps of the pet hospital, with smaller versions of the pets drawn on a 3 × 3 inch matrix

Description

For this activity, children search the pet hospital for the missing pets. Before the activity, the teacher positions the pets in various rooms of the hospital. Construction-paper doors, which are mounted to the poster board with tape, are closed to conceal the pets. The Pet Hospital game can be played in one of two ways, with verbal clues or with maps. When giving clues, the teacher should use precise location words. For example, he might say, "The fish is on the middle floor in the room in the center." Children can then instruct the teacher as to which door to open. If the doors are different colors, this may be easier to convey.

For the map version of the game, the teacher distributes premade maps that show the locations of the pets. The children must translate the smaller replicas of the hospital into the larger poster-board version in order to locate the pets. The teacher can point to various doors and ask the children which pet they think is in that room. Older preschool and kindergarten children may want to position the pets themselves and create their own location maps.

Math Discussions

Conversations surrounding this activity focus on positional terms, accompanied by pointing and other gestures that denote location. According to the research already discussed, teachers can expect children to have difficulty with the terms *right* and *left*; however, with appropriate scaffolding, use of the terms may help children begin to understand and use "right" and "left" themselves. Teachers can refer to the floors as "top," "middle," and "bottom." If an **ordinal number** is used to label each floor, the bottom floor should be identified as the *first* floor, the middle floor as *second*, and the top floor as *third*. Later, in school, when children locate positions on a graph, rows will be numbered starting with the horizontal axis, so labeling the bottom floor as first replicates that concept. It also mirrors labeling in the real world, in which the ground floor of a building is usually termed the first floor.

Throughout this activity, children should be encouraged to work together to solve the location problems. This gets everyone involved in the conversation. If the children use maps and direct the teacher to open a door that proves to be incorrect, the teacher can ask them to identify the animal that is in that particular room, find that animal on their maps, and then give her directions for which way to move in order to locate the desired pet.

Geometry Throughout the Curriculum

Geometry is already embedded in every area of the classroom, so each of these areas provides instructional opportunities. The more experiences children have with geometry in meaningful contexts, the more they can construct and solidify foundational concepts. Several examples of geometry integrated throughout the curriculum have already been described: spatial relationships modeled during music; the *Class Shape Big Book*, which includes photographs taken throughout the classroom; the Magnificent Shape Quilt, which is constructed in the art area; and the Where Are Our Friends activity, which takes place in the gross-motor or outdoor area. In addition, the importance of geometry in the block and manipulative areas has already been highlighted. More examples of geometry integrated throughout the curriculum follow.

Children's Literature

Many children's books relate directly to geometry, such as Tana Hoban's *Shapes, Shapes, Shapes* (1996). Less well known may be the wide selection of books related to quilting traditions, which by their nature emphasize shapes. Some examples are Ann Jonas, *The Quilt* (1994), a children's favorite that features an African American girl, a quilt with many shapes, and a lost toy; Georgia Guback, *Luka's Quilt* (1994), which highlights Hawaiian quilt traditions and their beautiful symmetry; Harriet Ziefert, *Before I Was Born* (1989), in which parents make a quilt for an expected baby; and Valerie Flourney, *The Patchwork Quilt* (1985), in which a young girl helps her grandmother finish a quilt.

Snack

How can children decompose three-dimensional solids? Snack time provides a solution. Many foods take the form of standard geometric solids: large marshmallows (cylinders), grapes (elliptical solids), cheese sticks (rectangular solids), melon balls (spheres), and cheese cubes (cubes) are some examples. During snack time, children can use plastic knives to slice their foods and examine the cut surface, or face, to see what shape it is. This could be an activity that takes place regularly throughout the year, whenever appropriate foods are served. Over the course of time, children can make observations and comparisons. For example, if a cheese cube is sliced straight down, the cut edge is a square; however, if it is cut diagonally, although the cut edge is still a rectangle, it is no longer a square. If a grape is cut in half across the short direction, the cut edge is circular; however, if it is cut across the long direction, the cut edge is elliptical.

Outside—Bubbles

Geometric bubble wands take geometric exploration to a new level. Their frames form the edges, or outlines, of geometric solids, which are revealed when children dip them into bubble solution. Often, the bubble film creates a secondary shape in the middle of the bubble wand. For example, the cube forms a clearly visible square in the center. Children also notice the lines moving from the corners of the cube to the corners of the central shape. Geometric bubble wands are commercially available or can be made with pipe cleaners.

 ## *Art Area*

A tremendous amount of geometry unfolds in the art area. Children can create imprints with geometric shapes, either with paint on paper or by pressing the shapes into clay or play dough; create symmetrical fold-over paintings; trace shape templates to combine or overlap shapes; and use shapes to create a variety of collage forms. Children often draw shapes as they move into representational art, and many renowned artists have a strong geometric component to their work. Children can examine examples of this artwork, discuss the shapes and how they are used, and then create their own artwork that is modeled on the artist's style. Some examples of artists with strong geometric content, whose works are readily available in books and on Internet sites, are Charley Harper, graphic nature artist; Frank Lloyd Wright, known for his drawings as well as his architecture; and Paul Klee, Swiss painter and graphic artist.

 ## *Dramatic Play*

The dramatic play area is an excellent place to focus on modeling positional terms. Here are some examples.

- *Put this corn inside the pan on the top of the stove.*
- *Should we put the cups on the top, middle, or bottom shelf of the cupboard?*
- *Here's our placemat to set the table. The fork goes on the left side of the plate, the knife goes on the right side of the plate, and the spoon is next to the knife.*
- *Can you put this fruit basket on the shelf below the vegetable basket?*
- *I can't find the salt shaker. Is it behind the milk bottle?*

UNIVERSAL DESIGN FOR SUPPORTING GEOMETRY

Many preschool and kindergarten children are identified with speech and language delays. They may need more experiences and time to develop an understanding of positional words. In fact, use of positional terms may be a specific individualized education program (IEP) goal for some children. Some of the activities described in this chapter, including incorporating positional words into a familiar song while modeling the actions, were originally designed to meet individual goals for particular children. As is often the case, the activities proved helpful to all children and, therefore, became part of the regular curriculum.

Some children with sensory integration problems may have difficulty describing the physical characteristics of objects or identifying them by feel. Because they often rely on visual characteristics for help, they may need to use visual clues for some of the activities in this chapter, such as the Grab Bag game. Experiences that combine visual, tactile, and verbal input may be helpful for children who are struggling to coordinate the senses.

Children with cognitive delays may be in a previsual stage of geometric reasoning in preschool and kindergarten. Whereas they may experiment with shapes, they may still have difficulty identifying them. Although many of the activities in this chapter are designed to move children into the descriptive level of geometric reasoning, they also support learning at the visualization level. Therefore, they accommodate all children. Teachers should, as always, vary their language and scaffolding on the basis of the level of thinking of individual children.

FOCUSING ON MATHEMATICAL PROCESSES

- *Problem Solving*—For preschool and kindergarten children, problems in the Geometry area often occur during play. They may have difficulty positioning a particular shape in a puzzle, creating a stable block structure, or locating a desired object in the classroom. Through experience, children learn to visualize how shapes look in various positions and develop strategies for solving puzzles that involve transformations. By experimenting, they learn to construct balanced and stable structures. Children also learn the meaning of positional terms and may mentally model the steps necessary to retrieve an object.

- *Reasoning and Proof*—Children use this process standard when they perform transformations and wonder whether the original shape is still the same. They may move the shape back and forth between the original and new positions to verify that, although the shape may look different in its new position, the shape itself has not changed. Reasoning and proof are also necessary when children group shapes on the basis of attributes. They may need to explain to their peers why a shape belongs, or does not belong, in a particular group.

- *Communication*—Communication is emphasized throughout the geometry curriculum as children discuss the labels and properties of various shapes. It is also an important component in specifying location (as opposed to merely pointing) and in describing spatial relationships.

- *Connections*—There are strong connections between algebra and geometry, because both involve patterning relationships, including symmetry. In addition, geometry and algebra both include classification and hierarchical organization. In geometry, children begin by grouping shapes that are the same, and later classify them according to number of sides and other attributes. Geometry is also connected to number, as children quantify the components of shapes, such as number of sides and points.

- *Representation*—As always, multiple ways of representation are important in mathematical learning. With geometry, children can represent geometric shapes by drawing them, connecting manipulative pieces to form shapes, or creating imprints in clay or play dough. Composition and decomposition of shapes can be represented with manipulative toys or shape collages. Children can represent transformations by moving shapes, as when they work puzzles, or through artwork, as in the Shape Aliens activity.

ASSESSING GEOMETRIC LEARNING

Work samples, photographs, anecdotal notes, and teacher-designed checklists are good ways to document learning in the Geometry area. In many of the activities in this chapter, children create a product that demonstrates their understanding. Shape Aliens, the Class Shape Big Book, Where Are Our Friends, and activities in the art area are examples. When teachers are finished displaying these projects, the work samples or copies of the work can be placed in an assessment portfolio. To record concepts associated with block building, digital photographs are particularly useful.

Anecdotal notes are helpful for assessing children's use of manipulative materials and their understanding and use of location words. The teacher should record exactly what the child says. On the basis of this information, teachers can emphasize particular terms and concepts with individual children or with the group. Anecdotal notes also capture children's geometric reasoning, which other types of documentation may not record.

Teachers may wish to create a checklist to document visualization concepts. They may need to know what shapes children can identify, whether or not children can identify shapes in various positions, and whether children can label shapes in nontraditional representations, such as triangles that are not equilateral or isosceles. This type of documentation is quick and easy to keep.

SUMMARY

This chapter emphasizes the four components of the geometry curriculum targeted in the NCTM Geometry standard: 1) analyzing the properties of two- and three-dimensional shapes; 2) specifying location and spatial relationships; 3) using transformations and symmetry; and 4) solving mathematical problems through visualization, spatial reasoning, and geometric modeling. Although many preschool and kindergarten children can name simple shapes, they often do not recognize them when they are in different orientations or in a less typical form. Spatial visualization is the ability to mentally produce and manipulate visual images of two- and three-dimensional objects. Whereas many children can visualize objects, being able to mentally move them takes longer to develop.

Dutch educator Pierre van Hiele has proposed a series of thinking levels that children progress through as they develop their understanding of geometric forms. The first two levels are the most relevant for preschool and kindergarten teachers. At Level 1, the Visual Level, children judge a shape according to its appearance. They often have a prototype in mind and may reject any shape that does not match that image. Children at Level 2, the Descriptive, or Analytic Level, begin to use language to describe the properties of shapes.

Many geometric, manipulative materials are available for children to use individually or in small groups. The block area is particularly important for the development of geometric concepts, including the properties of three-dimensional objects, balance, and symmetry. In addition, teachers can develop a range of activities for both small and large groups. Geometry should be regularly incorporated into the curriculum, through both planned activities and incidental learning opportunities throughout the classroom.

ON YOUR OWN

- Make a list of all of the geometry concepts you could emphasize on the playground or outside area. Consider lines, shapes, transformations, location, and spatial relationships.
- Create an "I Spy" game that uses positional terms.
- Make up a geometry song to sing with young children.

REFERENCES

Bowerman, M. (1996). Learning how to structure space for language: A crosslinguistic perspective. In P. Bloom, M.A. Peterson, L. Nadel, & M.F. Garrett (Eds.), *Language and space* (pp. 385–436). Cambridge, MA: MIT Press.

Clements, D.H. (2004). Geometric and spatial thinking in early childhood education. In D.H. Clements, J. Sarama, & A.M. DiBiase (Eds.), *Engaging young children in mathematics: Standards for early childhood mathematics education* (pp. 267–297). Mahwah, NJ: Erlbaum.

Clements, D.H., Battista, M.T., Sarama, J., & Swaminathan, S. (1997). Development of students' spatial thinking in a unit on geometric motions and area. *The Elementary School Journal, 98,* 171–186.

Clements, D.H., & Sarama, J. (2007). Early mathematics learning. In F.K. Lester, Jr. (Ed.), *Second handbook of research on mathematics teaching and learning* (p. 462). Reston, VA: National Council of Teachers of Mathematics.

Clements, D.H., Swaminathan, S., Hannibal, M.A.Z., & Sarama, J. (1999). Young children's concepts of shape. *Journal for Research in Mathematics Education, 30,* 192–212.

DeLoache, J.S. (1987). Rapid change in the symbolic functioning of young children. *Science, 238,* 1556–1557.

Flourney, V. (1985). *The patchwork quilt.* New York: Dial Books.

Froebel, F., & Jarvis, J. (1985). *The education of man.* New York: Augustus M. Kelley.

Guback, G. (1994). *Luka's quilt.* New York: Greenwillow Books.

Haith, M.M., & Benson, J.B. (1998). Infant cognition. In W. Damon, D. Kuhn., & S. Siegler (Eds.), *Handbook of child psychology* (5th ed.) (Vol. 2, pp. 199–254). New York: Wiley.

Hoban, T. (1996). *Shapes, shapes, shapes.* New York: Greenwillow Books.

Jonas, A. (1994). *The quilt.* New York: Puffin Books.

Liben, L.S., & Yekel, C.A. (1996). Preschoolers' understanding of plan and oblique maps: The role of geometric and representational correspondence. *Child Development, 67*(6), 2780–2796.

National Council of Teachers of Mathematics. (2006). *Curriculum focal points.* Reston, VA: Author.

National Council of Teachers of Mathematics. (2000). *Curriculum and evaluation standards for school mathematics.* Reston, VA: Author.

Piaget, J., & Inhelder, B. (1967). *The child's conception of space.* New York: W.W. Norton.

Piaget, J., & Inhelder, B. (1969). *The psychology of the child.* New York: Basic Books.

Rosser, R.A., Lane, S., & Mazzeo, J. (1988). Order and acquisition of related geometric competencies in young children. *Child Study Journal, 18,* 75–90.

van Hiele, P.M. (1986). *Structure and insight: A theory of mathematics education.* Orlando, FL: Academic Press.

van Hiele, P.M. (1999, February). Developing geometric thinking through activities that begin with play. *Teaching Children Mathematics, 6,* 310–13.

Wheatley, G.G. (1990). Spatial sense and mathematical learning. *Arithmetic Teacher, 37*(6), 10–11.

Williford, H.J. (1972). A study of transformational geometry instruction in the primary grades. *Journal for Research in Mathematics Education, 3,* 260–271.

Ziefert, H. (1989). *Before I was born.* New York: Knopf Books.

6

Developing Measurement Concepts

The body can provide you with the elements you need for measuring: fingers, an open hand or a fist, a forearm, a leg, and even your head, placed successively in a straight line to join two distant points. . . .

—Reggio Children (1997, p. 23)

Measurement is the third content area designated as a focus for the curriculum in the preschool, kindergarten, and primary grades (NCTM, 2006). Introductory and continuing experiences with measurement in preschool and kindergarten are considered important because they help children build foundational concepts that connect to later learning in the primary grades and beyond. Measurement, Number and Operations, and Geometry are strongly interrelated content areas. As children begin to construct an understanding of the sequential use of units when measuring, number is necessary to quantify, analyze, and compare those units. Geometry encompasses concepts of spatial orientation and location, and measurement is a mechanism for specifying distance between relative positions.

From infancy, children begin learning about measurement. As babies begin to focus on objects in their world and, subsequently, to reach for them, they begin to develop initial concepts of distance. With experience, they make finer and finer estimations of distance. Many a parent has been chagrined to find a toddler piling up furniture to grab a toy or cookie that is out of reach. Big and little, tall and short, and near and far are some of the first measurement comparisons that young children make.

THE MEASUREMENT STANDARD

There are two components to the Measurement standard (NCTM, 2000): The first deals with the measurable attributes of objects, and the second with appropriate tools and techniques. Expectations for the first component of the Preschool to Grade 2 grade band are that students

117

will compare and order objects according to attributes such as length, volume, weight, area, and time. In addition, they should select appropriate nonstandard and standard units and tools on the basis of the attribute that they wish to measure. Expectations for the second component are that students will use repetitions, or multiple copies, of a unit and develop mental images of common measures to use as reference points, such as the approximate length of a foot.

Even very young children begin to learn that objects have attributes which can be measured. Their first experiences come from their daily lives. Their height and weight are measured regularly on visits to the doctor; they discover that a big bucket holds more sand than a little bucket, and is also harder to lift; the top dresser drawer is too far to reach, but the bottom two drawers are not; big sister says she can run faster, and proves it. Measurements such as these can be made at the global, or perceptual, level, and children begin to learn the terminology that accompanies these comparisons: taller/shorter, heavier/lighter, farther/nearer, faster/slower, and so forth.

Children also begin to notice a second concept related to measurement, through their daily experiences—people use special tools to quantify measurements. For example, the nurse (or Mom or Dad) uses a scale to measure weight, and (in the United States) reports the amount in pounds. Mom uses a clock or her watch to determine when to leave for the bus stop, and talks about minutes or hours. Dad uses a yardstick to measure how high to hang a picture, and says the lines and numbers on it represent inches. In the classroom, children demonstrate this knowledge through their play. For example, they may draw a watch dial and tape it to their wrist, or they may place paper next to an object that they want to measure and draw slash marks and squiggles to replicate a ruler or yard stick.

The idea of a unit of measure that can be counted (iterated) to produce a measured result takes a long time to develop. At first, children usually select nonstandard units, such as the body parts that children in Italy used in the quotation at the beginning of this chapter. Shoes, blocks, chairs, crayons, or whatever is available may be pressed into service to measure length. The contents of smaller cups may be poured into larger ones and counted to give a crude measure of volume. Units of weight and time are more difficult to understand, because they cannot be visualized in the same manner as length and volume.

When children begin using a unit to measure length, measurement errors are typical. Children do not understand the need to place the units adjacent to one another in order to get an accurate measurement; therefore, gaps of varying sizes between the units are common. It is only through experience that children begin to think about this discrepancy, particularly if several children are measuring the same object and communicating different results. Children may also place the measurement units in a wavy, rather than straight, row, which also confounds their results.

The examples that follow illustrate ways in which young children may explore measurement.

EXAMPLE 6.1

Sophie was determined to reach the ceiling of her preschool classroom. She sat on the floor of the block area, carefully connecting transparent cylinders. Periodically, she lifted her cylinder column and reached toward the ceiling. After 45 minutes of intense work, Sophie extended her column once more toward the ceiling and touched a ceiling tile. "I did it," Sophie announced happily. "I touched the ceiling." Her teacher helped her count how many cylinders it had taken to span that immense (to Sophie) distance.

In this vignette, a 4-year-old child tackles the difficult challenge of measuring a distance that she cannot reach. Interlocking manipulative materials provide the perfect tool to help her reach this goal. Because the units hook together, there is no gap between them. This is helpful when children are learning to measure linear distance and essential when they must "grow" their measuring tool, as in Sophie's case. Teachers should remember to suggest materials such as these for in-class measurements.

EXAMPLE 6.2

Raven and Mia had decided that they would be princesses in the class movie. Each child got to choose a role, and the teacher would videotape the movie as he narrated the story that the children had written. This was costume-making week. The teacher had provided several yards of tulle, a netted fabric, and Raven and Mia thought that this would be perfect for their gowns. Raven held up the fabric to Mia and asked her to hold it. Then, Raven cut across the tulle at Mia's ankles. "There," Raven said, "now your dress will be long enough. Now you do that for me." Soon, both girls had an appropriate length of fabric to create their gowns.

One of the easiest ways for children to measure is to make a direct comparison with another object. In this case, it was obvious to Raven that the dresses should extend from their shoulders to their ankles. Using body length as her reference, she created a length of fabric that approximated her model.

EXAMPLE 6.3

Mark received a set of large, plastic nesting blocks for his first birthday. There were many ways to play with them. Mark could stack them up from largest to smallest and knock them down, which he liked to do. He also spent a lot of time putting blocks inside other blocks and then taking them apart. Sometimes, the block Mark wanted to place inside another block would not fit, and he had to try it again. Then Mark discovered a new way to use his blocks. He put one on his head and laughed as his dad called it a hat. Then he put a block on each foot and trudged around as if the blocks were boots. The next time Dad glanced up from his paper, he found that Mark had found yet another use for his blocks. He had squeezed himself into the largest block and was happily sitting in it.

Mark's explorations illustrate how young children form initial size relationships. Using objects in their environment, including their own bodies, they diligently experiment to determine what will fit inside what. Many a parent has wryly observed that the box that a present came in was more popular than the present itself. Stacking, nesting, and comparing blocks to body parts are all part of children's early measurement explorations.

EXAMPLE 6.4

It was an exciting day for the kindergarten. The children had been studying sea animals in class and discovered that the biggest animal in the world is the blue whale. Before school, the teacher taped the outline of a blue whale to the floor of the cafeteria. After the opening group, the children walked to the cafeteria to measure themselves against the blue whale. One by one, the children lay down on the floor inside the outline, with the feet of one child touching the head of the next child. "Oh, no," said the teacher. "We've run out of children, and we still haven't reached the end of the whale." The teacher carefully marked the end of the last child's

---

(continued)

> **EXAMPLE 6.4** *(continued)*
>
> *feet. Then the children stood up and gazed at the immense length of the whale.*
>   *"It's longer than all of us," Nilani said.*
>   *"How many children long is that?" asked the teacher.*
>   *"We have 25 children in our class," answered Sammy, "and everyone is here today, so it's bigger than 25 children."*
>   *"I think Miss Williams should lie down and measure, too," said Mindy, "because we didn't get to the end of the whale."*
>   *The teacher agreed and lay down with her head touching the tape mark for the last child. "Do my feet touch the end of the whale?" she asked.*
>   *"Almost!" Sammy shouted.*
>   *Miss Williams stood up. "So how long is the whale?" she asked.*
>   *"It's 25 children and one grown-up," said Mindy.*
>   *"And a little bit more," added Sammy.*

This experience capitalizes on the natural interest of young children to compare themselves to something of interest, such as the biggest animal in the world. The teacher makes the 100-foot length of the blue whale more understandable to the children by creating its outline on the floor so that they can visualize it. The unit of measure, their own body length, is not a standard unit, but it is something that they can relate to. The next time the children read about the blue whale, they can envision their entire class and remember that the whale was longer than all of them, plus their teacher. The approximate height of a kindergarten child is a useful reference point for them.

> **EXAMPLE 6.5**
>
> *Sean was busy comparing objects on the balance scale in the science area of his classroom. He put a pumpkin gourd on one side of the scale and a rock on the other side. The tray with the rock moved down while the pumpkin moved up. "Which one is heavier?" asked Sean's teacher. "The pumpkin," Sean happily replied. "Why do you think the pumpkin is heavier?" the teacher asked. "Because it goes up," Sean confidently replied.*

This scenario with Sean is familiar to teachers who work with young children and illustrates children's confusion about the concept of weight. Although children can often accurately compare objects that are heavy and light, when the objects are placed on a scale, children frequently believe that the side that goes up is the heavy one. This seems to be another instance in which children's perceptions fool them. Children are used to comparing the size of objects, and they learn to associate big with heavy. Because bigger things are taller (extend higher from the floor or table), children may assume that they are also heavier. Teachers can encourage children to handle objects of differing weights and then compare what happens when they are placed on a scale; nevertheless, teachers should expect some children to need considerable time before they can relate the movement of the scale to the weight of the object.

THE DEVELOPMENT OF MEASUREMENT CONCEPTS

The development of measurement concepts begins in infancy, but continues to develop over many years (Piaget & Inhelder, 1967). Even 9-month-old babies have been documented as discriminating sequences of size (Brannon, 2002). Preschool children have several strategies for judging relative size. First, they use perceptual comparisons when both objects are present, such as a big ball versus a small one. Second, they can use a remembered object as a norm for comparison of other objects. For example, a child who has a German shepherd at home may call a Scotty a little dog. Finally, preschool children can use functionality to make size decisions, such as selecting an appropriately sized hat for a doll (Gelman & Ebeling, 1989). However, preschool children do not distinguish between continuous and **discrete quantity** (Piaget, Inhelder, & Szeminska, 1960). Therefore, equal sharing is based on the number of items, such as pieces of cracker, rather than the total amount. As an example, a 2-year-old who was given a cup of juice demanded more from his mother, before even beginning to drink the juice. She nonchalantly took a second cup, poured part of the original juice into it, and handed it to the child, who was now content. Even though the amount of juice had not changed, from the child's perspective, it had grown from one glass to two.

Measuring Length

Much of the research related to the development of measurement concepts in young children is in the area of **linear measurement.** Although children can learn measurement procedures, research indicates that underlying measurement concepts are much more difficult for children to understand and take years to develop. Piaget described measurement of length as the dividing of a continuous space into interlocking segments (Piaget & Inhelder, 1969a). One of the segments must then be used as a unit and applied to the entire space, without overlapping, to determine the measurement. The child must synthesize two concepts: 1) the displacement of space by the segments and 2) the reunification of the units through addition. This conceptualization takes years to fully develop.

When comparing the lengths of objects, preschool and kindergarten children are often confused by the placement of the objects. Although they can tell that two identical rods placed next to one another are the same length, if one rod is moved so that its tip extends beyond the other, many 4- to 6-year-olds believe that the protruding rod is longer. By age 7, most children can conserve length and do not make this mistake (Inhelder, Sinclair, & Bovet, 1974).

Most preschool and kindergarten children also do not understand that a straight line is the shortest distance between two points. When shown two routes between the same two points, one straight and the other curved, most preschool children and about half of kindergarten-age children believe that both routes are the same length (Fabricius & Wellman, 1993; Miller & Baillargeon, 1990; Piaget, Inhelder, & Szeminska, 1960).

Despite their limitations in reasoning, young children do successfully invent strategies to solve measurement problems, as demonstrated by the children in Reggio Emilia, Italy (Reggio Children, 1997). The task of these 5-year-old children was to provide the local carpenter with the measurements of a table in their classroom so that he could build an identical one. The children began by laying their fingers on the table and counting. Subsequently, they tried fists, hand spreads, legs, a kitchen ladle, and a book. Eventually, the children constructed their own "meter" sticks, complete with numerals and hash marks, only to discover that each meter stick gave a different measurement for the table. The children learned through these experiences, and eventually succeeded in measuring the table with one of the children's shoes, and then with an actual meter stick. This accounting shows that when children can work together on a project of intense interest, they can surpass learning expectations.

In a much less involved project, three preschool children in the author's class were observed measuring the length of a "snake" that they had constructed by connecting interlocking blocks. The children lined up chairs, front touching back, and placed their snake along the seats of the chairs. They then counted the chairs to determine the length of the snake. Several important measurement concepts were demonstrated. First, the children used a standard-sized chair for their unit of measure. Second, they placed the chairs so that they touched one another. Third, they aligned the end of the snake with the end of the first chair. Fourth, they counted the units to determine the length of the snake. All of this was done without adult involvement. One conceptual error that the children demonstrated was not stretching the snake out into a straight line; instead, it had a small curve, possibly so that it would fit on the row of chairs without extending beyond the seat of the last chair.

Measuring Area

When measuring area, children have to contend with two dimensions, length and width. Thinking about two attributes simultaneously, as when sorting collections according to both color and size, is difficult for young children. Some research indicates that preschool children focus on only one dimension, such as length, when examining area (Bausano & Jeffrey, 1975; Piaget, Inhelder, & Szeminska, 1960; Raven & Gelman, 1984; Sena & Smith, 1990). However, if preschool- and kindergarten-age children have the opportunity to play with two-dimensional materials before being tested, they are more likely to use strategies consistent with multiplicative rules, or consideration of length and width (Wolf, 1995). By 5 or 6 years of age, children are able to manipulate regions by placing two-dimensional forms on top of them (Yuzawa, Bart, Kinne, Sukemune, & Kataoa, 1999).

Some inkling as to young children's reasoning about area comes from Piaget's cows-in-the-field **conservation** problem (Piaget, Inhelder, & Szeminska, 1960). Two identical sheets of green paper, representing two pastures, each with a toy cow on it, are placed before the child. Identical blocks are used for barns. First, one block is placed on each paper field. Children generally agree that both cows still have the same amount of grass to eat. Then, a second barn is added to each field; however, on one field the barns are separated from one another, whereas on the other field they are adjacent. Children younger than the ages of 7–8 years usually believe that the cow in the field with contiguous barns has more grass to eat than the cow in the field with two barns that are separated. Even though the barns cover the same area, nonconservers view the pasture with adjacent barns as having only one barn, whereas the other pasture has two barns and, therefore, less grass for the cow to eat.

Measuring Volume

In determining capacity, or how much a container will hold, young children once again rely on their perceptions. Piaget demonstrated that most children younger than 7 years of age believe that the amount of liquid contained in a glass can change when it is poured into another glass with a different configuration (Piaget, 1952; Piaget & Inhelder, 1969a). When children are shown two glasses, each with the same amount of liquid, and one of the glasses is poured into a tall, thin glass, most 4- to 6-year-old children believe that the tall glass now contains more water than the other glass. Why do children make this error in reasoning? Piaget's view is that children focus only on the water levels in the short and tall water glasses, and they do not think about the transformation, or the transference of water from one glass to another. In addition, they are not able to reverse their thinking and reason that the water could easily be poured back into the original glass.

Although conservation cannot be taught and, therefore, should not be approached as an instructional goal, research indicates that social interaction helps children move forward in their understanding of conservation concepts. Kamii (2000) describes several studies in which children who were nonconservers were grouped with either conservers or other nonconservers and encouraged to interact on conservation tasks. Many of the children who were grouped with conservers were shown to have moved forward in their understanding of conservation concepts at posttest, whereas almost none of the children who were grouped with other nonconservers showed this progress. Kamii stresses that the difference was in the dialogues. Conservers gave logical explanations accompanied by demonstrations to their nonconserving peers, and this appeared to influence the reasoning of the nonconservers.

In the classroom, teachers are likely to observe similar conservation issues when children compare the volumes of various containers in the water table. Children typically reason that tall, thin containers have more capacity than short, squat ones. Longtime teacher Angela Giglio Andrews noticed this phenomenon in her kindergarten classroom when the children explored containers in the rice table (Andrews & Trafton, 2002). They were convinced that a tall ketchup bottle would hold more rice than a squat peanut butter jar. After experimenting by pouring scoops of rice into both jars, many of the children were amazed to discover that the peanut butter jar held more rice. By the end of kindergarten, after having numerous opportunities to measure the capacity of containers in the rice table, the children were able to arrange in order by volume, and with only one small error, a series of seven containers with widely varying appearances.

Measuring Mass

When comparing the masses of objects, children have conservation issues which are similar to the issues that they have when measuring length and volume. If they have two piles of play dough that they judge to be the same size, and one of the piles is stretched out, most preschool and many kindergarten children believe that the pile which is longer now has more play dough. Children who can conserve mass readily explain that they know that the two piles are still the same because no play dough was added or taken away. They may demonstrate that when they push the play dough back to its original position, it looks the same as the other pile.

DESIGNING THE MEASUREMENT CURRICULUM

The focus of the Measurement standard for young children is on recognizing measurable attributes of objects and developing and utilizing tools and techniques that are appropriate for measuring these attributes. Because the direct comparison of objects is a precursor to more detailed measurement, teachers should provide many opportunities for children to compare the measurable attributes of materials in the classroom. The block area is an excellent source of comparisons, because unit blocks are constructed to show proportional lengths. A double-unit block is twice as long as a unit block, and a half-unit block is half as long as a unit block. Children can be encouraged to move blocks of the same length into various positions and then recompare them so that they can see that the blocks have not changed size. For example, children can experiment with blocks judged to be the same and different lengths, to judge the distance needed to bridge two vertical blocks. All areas of the classroom provide opportunities for making size comparisons: pans in the dramatic play area, pencils in the art area, sizes of books in the literacy area, and so forth.

Children also need many opportunities to consider the use of units for measuring. Although adults may be inclined to tell children how to use measuring units so that they do not make mistakes, children need the opportunity to experiment and determine the function of units for themselves. Through observing and discussing the discrepancies in their measurements, children construct foundational measurement concepts. Rather than memorizing a rule, they understand *why* the units need to touch one another (but not overlap) and the consequences if they do not.

Finally, although linear measurement may be easier for young children to visualize than area, volume, or mass, children need foundational experiences with all of these attributes. Even though a mature understanding of measurement is not expected of children for many years, children do make progress in their conceptualization of measurement during pre-school and kindergarten. These foundations are important for later learning.

Math Talk

Conversations related to measurement should center around the developmental thinking of the children and concepts that lie on the immediate horizon of their learning progression. Because young children begin by making direct comparisons of objects, teachers should support this foundational concept. During clean-up time, the teacher might ask the children to line up the farm animals and trucks on the shelf, according to size. On occasion, the books might be arranged on the shelf according to height, or measuring cups in the dramatic play area aligned according to size. Teachers can also focus on opportunities throughout the day to draw attention to measurement comparisons. A few examples follow:

- *Are all the carrot sticks on your plate exactly the same length? I'm going to line up mine and compare them.*

- *Susie, we want to make a bridge between these two blocks. Can you find a block that will stretch across both of them?*

- *Lots of people brought in leaves for the science table today. Let's put them in order by size.*

- *Our extra-clothes bin is a mess. Who can help me stack the shirts and pants according to size?*

Preschool and kindergarten children can also compare objects to mental norms that they have constructed for similar objects. On the basis of those norms, children might comment, "That beach ball is super big," or "The baby bunny is really little." Teachers can help children extend their comparative reasoning by playing "I Spy" or "I'm thinking of . . ." games that relate to measurement attributes, such as the following:

- *I spy something that's round like the clock, but much bigger.*

- *I spy a very big book.*

- *I'm thinking of an animal that's smaller than a cat.*

- *I'm thinking of an animal that walks very slowly.*

- *I'm thinking of a plant that's taller than a flower but shorter than a tree.*

During the preschool and kindergarten years, young children begin to construct the concept of a unit of measure. Teachers can begin to interject comments that point children in that direction. Here are some suggestions:

- *I want to make my block tower as tall as yours. How many more blocks do I need? My tower is 1, 2, 3, 4, 5 blocks high.*

- *I think our dramatic play table would look nice with a tablecloth. If you can figure out how big the tablecloth should be, I'll make one. Check the manipulative area for something you could use to measure the table.*

- *Look how tall Olivia's building is! Let's measure it with unit blocks and then with double-units.*

- *Jason, how many straw pieces did it take to make your bracelet? Your wrist is about 7 straw pieces around.*

Individual or Small-Group Activities

Children often seem intensely interested in linking together manipulative objects, such as snap-together blocks, to make the longest strand possible. As in the earlier example of children who made a snake, they may then contemplate just how long their creation has become and seek some method of measurement comparison. Collaboration with friends is often a key component of these spontaneous measurement experiences. For this reason, small-group experiences are an excellent venue for planning and encouraging measurement.

Whereas measurement activities may evolve naturally from the play of young children, they are more likely to occur when teachers intentionally plan activities that encourage children to make size comparisons and to **unitize.** Teachers can then be prepared to offer comments, ask questions, or model measurement concepts through play. For example, if a fishing pond is set up in the sensory table or dramatic play area, the teacher can ask who caught the smallest fish or suggest that they measure the various fish with interlocking cubes. The activities described in this section are specifically designed to elicit measurement opportunities.

Linear Measurement

Linear measurement encompasses two concepts: 1) length, or the span between the starting and ending points of an object, and 2) distance, or the space between two points. Children often use direct comparison to judge the lengths of two objects, but they can also use an intermediary object to measure either length or distance. For example, they could use a stick to mark the height of a plant which they are growing and then compare that mark to a friend's plant, or they might lay a block between two structures to determine what it would take to span the distance. In later preschool and kindergarten, children can begin to experiment with identifying a unit of measure and quantifying it. The activities described next are designed to encourage all three types of measurement so that children can conceptualize how the measurement types relate to one another. Recall the account in Chapter 2, in which children interacted with quantification games according to their level of understanding: global, one-to-one correspondence, or counting. The opportunity to observe peers at a higher level of thinking, combined with the conversations related to the play, helped move children forward in their thinking and solidified the concepts of all of the children. In much the same way, the measurement activities that follow allow children to participate at three levels of understanding: 1) direct comparison, which is the easiest form of measurement; 2) use of an intermediary object, which is more difficult; and 3) unitizing, which is the most complex of the concepts to understand.

ACTIVITY 6.1

Those Amazing Pumpkins

Materials

The following materials are needed for this activity:

- Several pumpkins of different sizes
- String, in two colors
- Yard stick and tape measures

Description

Children are fascinated with pumpkins, which are a popular autumn fixture in classrooms. Even young

preschoolers can arrange the pumpkins by size and apply appropriate labels, such as big and little. There are many measurement opportunities inherent in a pumpkin. In order to create interest in measurement, the teacher might begin by asking the children at group time whether they think the pumpkins are larger when measured from bottom to top or around the circumference. Predictions could be charted. Then, throughout the week, individuals or small groups of children can measure the pumpkin in both dimensions. Use of a different color of string for length versus circumference helps children distinguish between the two measurements. At the end of the week, children can compare the lengths of string for the pumpkins and determine whether the length or the circumference is the longest.

If all of the measurements are displayed together on a chart, some children may notice that they are not exactly the same. This can spark a discussion about why different groups recorded different measurements. Did the pumpkins change in size, or was there a difference in the way people measured them? Some children may want a more exact measurement of the pumpkins. The teacher can work with these children to compare the length of the string to a yardstick or measuring tape.

Math Discussions

There is much room for discussion at various stages of this activity. First, children can discuss why they think the length of a particular pumpkin will be more, less, or the same as its width. This conversation can be initiated during the large-group introduction to the activity and carried over into the individual and small-group measuring activities. As children complete their measurements, they can compare the strings that they used to measure the length versus the circumference of the pumpkins. Was one color of string longer each time, or did the results vary? Kindergarten children may want to record their discoveries in a journal. Teachers should also encourage children to compare the lengths of string they used for various pumpkins. This encourages measurement comparisons by the use of an intermediary object. The children may discover some interesting relationships, such as that the string for the circumference of the large pumpkin wraps twice around the medium pumpkin. Finally, children can be encouraged to measure the length of their strings with 1-inch cubes and a yardstick, and discuss their findings. Finding discrepancies in their measures may help children realize why the cubes need to be touching one another to get an accurate measure.

ACTIVITY 6.2

Shoe Store

Materials

The following materials are needed for this activity:

- Assortment of shoes in many different sizes

- Tag board, to create a pattern of each child's foot by tracing around the foot and cutting out the pattern (foot cutouts should be labeled with the child's name)

- Several 12-inch foot templates, made the same way

- Ruler, for comparison

- Note cards, to create shoe-size signs

Children love trying on shoes, so this dramatic play activity generates considerable interest. At first, children need time to play with the shoes. By trying on various shoes, children quickly discover that some are too big, and others may be too small. This is a good time to start talking about the relative size of the shoes. Children may decide to group the shoes by size, and some may recall that shoes in stores are arranged by numbered sizes. There are many possible outcomes to this activity, and they may change from day to day. Some children may want to order all the shoes by size. Others may decide to create signs for various shoe sizes. These could range from big, medium, and small to actual numbers to designate a size.

After children have had time to play in the shoe store for several days, the shoe template activity can be introduced. Preschool teachers can trace around the children's feet and cut out the templates. In kindergarten, teachers might group the children in pairs and have them trace each other's feet. Children can compare their foot templates to various shoes in the shoe store and to the templates of one another. They may also want to compare their foot to the teacher's foot. During the second week of the activity, the 12-inch foot templates can be introduced, along with a ruler, if desired. Children can measure distances in the room with the standard-foot templates and their own foot templates, and compare the results. In this way, children begin to consider the reverse relationship between the size of the measuring unit and the number of iterations required to measure something. In children's terms, it takes more little feet than big feet to measure something. This is counterintuitive to young children because they associate something large with a bigger number.

Math Discussions

There should be much engaged conversation throughout this activity. At first, children will want to talk about the various sizes of shoes and whether or not they fit. There may be heated discussions, particularly in kindergarten, about how to determine the sizes for the shoe signs. Later, conversation will turn to comparing the sizes of everyone's feet, and finally, to using the shoe templates to make measurements.

Teachers have an important role in guiding these conversations. For example, they might encourage children to measure the same distance, first with their foot template and then by pacing it out, heel to toe. Is the distance the same, and if not, why not? Teachers can also encourage children to discuss why the same object may have a different measurement depending on whose foot template is used. This encourages children to think about the size of the unit that they are using to measure. Finally, children may want to measure their foot template by using various objects from the classroom, such as cubes, crayons, or their fingers.

ACTIVITY 6.3

Inchworms

Materials

The following materials are needed for this activity:

- A copy of the book *Inch by Inch,* by Leo Lionni (1960)
- Interlocking inchworms, commercially available or made by adhering Velcro circles to the sides of 1-inch cubes and gluing "worms," made from foam strips, to the tops
- Measurement fill-in strips, such as the following:

The airplane was 12 inches long and 12 inches wide.

| The _____ was _____ inches long. |

Description

Young children do not yet understand that the units used to measure length must be contiguous (touch each other), or the measurement will not be accurate. In this activity, interlocking toy inchworms give children experience in measuring length by using discrete units that connect, and therefore touch one another. The children's book *Inch by Inch* provides the inspiration. In the story, an inchworm convinces a robin not to eat it because its inch-long body is useful for measuring things. After the children have heard the story, the teacher can introduce the inchworm manipulative materials and encourage them to measure objects of interest. The fill-in measurement strips enable children to keep track of their measurements for later comparisons.

Math Discussions

As children conduct their measurement experiments, teachers will likely notice some conceptual errors. For example, some children may not start their measurement at the edge of the object. The teacher might ask whether it is important to measure the part that is not covered by the inchworms. Another tactic would be to have two children measure the same object and compare their results. Teachers may also notice that when children measure across a wide item, such as a table, their measurement tool may be positioned at a diagonal rather than parallel to the edge. In this case, the teacher might suggest that the children measure along the edge of the table with another set of worms, and then discuss how the two lines of interlocking worms compare.

Many young children are confounded when an object is placed in a different position. As an example, they may think that the length of a book changes when it is moved from a horizontal to a vertical position. For this reason, teachers should encourage children to take measurements of the same object in different positions. Be aware, though, that children may get different measurements due to errors that they make, such as not aligning the measurement tool with the bottom of the object. In this case, the teacher can serve as a memory of what was done in the previous measurement. For example, the teacher might say, "I noticed that when you measured the book lying flat on the table, you started the inchworms here (point to the position), but when you measured the book standing up, you started the worms at the very bottom of the book. Do you think that makes a difference?"

Measuring Area

Area measurement must take into account dimensions of both length and width. In a sense, the Shoe Store activity, as discussed in the previous section, was a measure of area; however, it was used to focus on linear measurements. The activity described next relates directly to area measurement.

ACTIVITY 6.4

Carpeting the House

Materials

The following materials are needed for this activity:

- Unit blocks
- Square tiles, 2¾ inches per side, cut from colored fun foam (available in craft stores)
- Digital camera (optional)

Description

The purpose of this activity is to provide foundational experiences related to the concept of area. Whereas most measurement activities in preschool and kindergarten deal with the concept of length, this activity focuses on the relationship between length and width.

Many children like to build houses in the block area, particularly if accessories, such as small toy people and dollhouse furniture, are available. In this activity, children can use foam tiles to carpet their house. The tiles are cut into 2¾-inch squares so that they will align with standard-unit block sizes. A single tile will cover one half-unit block, two tiles will cover a single-unit block, and four tiles will cover a double unit. Because children will use the unit blocks to form the perimeters of their houses, tiles of this size allow them to cover the created area. Many children have carpeting or floor tiles in their homes or schools; therefore, they are accustomed to the idea of a continuous floor covering.

Math Discussions

Children will need time to experiment with the floor tiles before much conversation can take place. Teachers should be patient and merely observe while this process unfolds. When children appear to have accomplished their purpose, teachers can talk to them about their goals. Some children may at first just place the tiles randomly inside their structure. Teachers should always begin by acknowledging what the child has done. In this case, the teacher might say, "I notice you put some blue tiles inside your building. They're softer than the blocks, aren't they?" At this point, the teacher can draw the child's attention to the floor of the block area and talk about how the carpet or tiles go from wall to wall, with no gaps. The teacher might say, "I have an idea. I'm going to build a house like yours and see whether I can use these tiles to make the floor." The teacher can then model laying the tiles in continuous rows across the length and width of the inside of the house. This may encourage children to also experiment with covering area.

Other children may immediately begin using the tiles to cover the floor area of their structure. In this case, teachers can ask how many tiles they needed. Children can also count how many rows they have in each direction and how many tiles are in each row.

Measuring Volume and Mass

Volume and mass each involve a **continuous quantity,** a concept of measurement that is more difficult for children to separate into units and to visualize than length. In addition, volume and mass are contained in three-dimensional space, in contrast to length, which varies in one dimension, and area, which varies in two. Containers that look very different, such as a tall, thin jar versus a broad, squat jar, can contain equivalent amounts. The mass of an object, such as a mound of clay, can take many different shapes without the amount of clay changing. For these reasons, measuring volume and mass pose particular challenges to children. Nevertheless, children are highly interested in experimenting with capacity. Guided explorations of materials during preschool and kindergarten can help them form foundational concepts for later, more advanced measurement. The two activities described next exemplify the types of experiences that teachers can plan to help children think about the measurement of volume and mass.

ACTIVITY 6.5

Container "Brainer"

Materials

The following materials are needed for this activity:

- Sensory table, or several plastic dishpans
- Assorted clear containers, in various sizes and shapes
- Several small scoops
- Several funnels
- Recording chart (optional)

Description

This activity is designed to help children visualize the various parameters that affect volume. Preschool and kindergarten children typically believe that tall containers automatically hold more than shorter, wider containers. Therefore, for this activity, teachers should select containers that confound that notion. Some taller containers should be included that have less capacity than shorter containers. At first, children will need time to randomly experiment with the materials, particularly in preschool. The teacher can then introduce the challenge of finding the containers that hold the most water. The idea of using the funnels and scoops to measure how much water goes into each container can be modeled, perhaps during group time. For recording purposes, a chart with a list of the children's names and a digital photo of each container can be placed near the activity. Children can add comments, such as "a lot," or measurement numbers, such as "5 scoops," to the chart. To facilitate later discussion, the bottles can be labeled with letters; however, numbered labeling should be avoided because children may confuse the numbers with a volume hierarchy or think that the numbers refer to the measurements.

This activity should be repeated, perhaps during the following week, using a dry material, such as sand or rice. Some children believe that the filler material affects the capacity of the containers, and this variation allows them to check that hypothesis.

Math Discussions

Conversation is an important component of this activity. Before the activity begins, teachers can ask the children to estimate the number of scoops that each container will hold. This can be done informally, just before children engage in the activity, or as a formal **estimation** activity at group time, with the estimates listed on chart paper.

Scaffolding during the activity will be critical for many children. In preschool, teachers can help children keep count of the number of scoops that they poor into various containers. Kindergarten children can be grouped in pairs or small groups so that they can help each other keep track of the counts. Teachers may also suggest that children pour the contents of particular containers into other containers to check their predictions. For example, the teacher may suggest that the sand in a tall container be poured into a shorter container to see whether it overflows.

It is important to remember that for some children, particularly preschoolers, this will be a general, introductory activity. It takes many experiences over time for children to develop measurement concepts related to volume. For younger children, the teacher's role may simply be to draw children's attention to close observation. For example, the teacher might say, "Tom poured the water from the tall bottle into this shorter bottle. Did it fill the short bottle all the way up?"

For older children, the charted measurements can be discussed as a group. There will undoubtedly be discrepancies, and children can speculate about why this would happen. The teacher may want to demonstrate some of the suggestions, such as what happens if the scoop is not filled to the top or if children keep counting after a container is filled. During the conversation, teachers can draw children's attention to a comparison of pairs of containers. They may notice that container C always holds more than container D. If this conversation is held after the experiments with water, the teacher might ask children to predict whether or not the same pattern will emerge when they use sand or rice.

Snake Adventure

Materials

The following materials are needed for this activity:

- Small container of play dough for each child

- Measurement tools, such as string, linking cubes, or standard measures (e.g., ruler, yard stick)

- Recording sheet (optional)

Discussion

Children who do not yet conserve mass, which includes most children in preschool and many in kindergarten, typically believe that the mass of an object changes when it is spread out or compressed. Therefore, they think that a lump of play dough that has been spread out now has more mass. In this activity, children experiment with stretching out play dough and then returning it to its original container.

Children frequently experiment with coiling by rolling play dough or clay back and forth across a flat surface with their hands. As they do this, the dough becomes longer and thinner. Children often announce that they have made a snake. This activity extends this natural focus of children by encouraging them to measure their snakes, speculate as to whether the play dough will still fit into its original container, and then see for themselves. There are various ways that children can measure their snakes. They can cut a piece of string that is as long as the snake, use interlocking cubes or other manipulative materials, or compare the snake to a standard measure. The activity thus involves both linear measurement and comparison of mass.

Children who are close to conserving, or understanding that no mass has been lost or gained by manipulating it, may be highly intrigued (or bothered) by their experiments. They may repeat the process over and over to see whether the results remain the same. Within a few days, they may have convinced themselves that the mass does not change. However, younger children who are not developmentally ready to conserve may not be bothered at all by the process, because they remain convinced that the amount of play dough actually changes as they manipulate it. Nevertheless, the activity gives them experience in manipulating a continuous quantity, some beginning experiences with measurement, and the opportunity to make focused observations accompanied by conversation.

Math Discussions

Once again, conversations should focus on predictions, observations, and discussions about the results of the experiments. Teachers may wish to begin by modeling the coiling technique, although play dough can also be stretched without coiling. Children can begin with estimates of how long they think they can make their snake (or stretch their play dough). They can then conduct and record their measurements. Finally, they can speculate as to whether the play dough will go back into its container, and then find out. Teachers' comments and questions guide this process. For example, the teacher might use tape to mark the child's measurement prediction on the table or floor. Teachers will certainly want to draw attention to how the width of the snake changes as it gets longer; in fact, this can be measured as well. Teachers can also interject ideas for other experiments; for instance, if half of the play dough in the container is used, will the snake be as long?

The research cited earlier in this chapter indicates that the dialogues among children are a critical component in increasing the understanding of nonconservers. For this reason, kindergarten teachers may want to group together children with various levels of understanding about measurement concepts. More advanced children can explain and demonstrate their reasoning, which adds to their own mathematical thinking while also encouraging their peers to think about possibilities that are new to them.

Measuring Speed and Time

Children typically do not understand the relationship between speed and time until about age 10 or 11 (Piaget & Inhelder, 1969b; Wadsworth, 1989). The formal relationship is expressed as velocity = speed/time. Younger children fail to consider whether or not two objects whose speed they are comparing started at the same time and followed the same paths. Nevertheless, young children are very interested in racing objects. The activity that follows focuses on physics. Children can initially rely on their visual perception to judge speed. Later, they can compare how speed relates to distance traveled and even can conduct general measures of elapsed time.

ACTIVITY 6.7

Ramp Races

Materials

The following materials are needed for this activity:

- ■ Two pieces of wood, 4 inches wide and 24 inches long, to form the racing tracks

- ■ Two unit blocks or pieces of wood 2 × 4 inches, of which one is about 5 inches long and the other about 2½ inches long

- ■ Two identical toy cars, small enough to roll freely down the track

Description

This activity allows children to compare the effect of slope on the speed of a rolling object. Two sizes of blocks are used to elevate the molding, thus creating inclines with different slopes. Children can roll the cars down the two slopes and determine which incline generates more speed. It is usually quickly apparent to children that the car on the steeper slope goes faster; however, children will want to replicate the experiment many times to assure themselves that this is always true. (In this sense, they are good scientists!)

Once children have had ample time to explore the ramps, tape lines can be placed on the floor at regular intervals, such as every 12 inches. Children can use these marks to quantify and compare how far each car rolls. They should discover that the car which moves faster also travels farther.

As a final aspect of the experiment, children can begin to measure the amount of time that it takes for the two cars to reach the bottom of the ramps. This can be done in two ways. First, if the teacher has access to a metronome, a small electronic device that produces clicks at regular speeds for musicians to follow, children can count the number of clicks that it takes for each car to move down the ramp. Another technique is for teachers and children to clap the steady beat in a familiar song, such as "Twinkle, Twinkle, Little Star." Galileo used this technique when measuring acceleration on inclines (Johnson, 2008). Once a steady beat has been established, the car can be released and the number of claps counted until the car reaches the end of the ramp.

Math Discussions

Each phase of this experimental process should elicit conversations among the children. They will initially be interested in comparing the speed of the cars on the two ramps. The teacher should guide them to accurately describe the ramps that they are referring to, such as "high versus low" or "tall versus short."

When children move to the linear measurement phase of the activity, discussion will likely revolve around which car moved the farthest. For the younger children, this will require a change in focus (away from just looking at speed) that may at first be difficult. Some children may simply reply in relative terms, such as, "This car went farther." Older children may decide to count the number of lines each car crosses. Teachers should guide the conversation so that children focus on the relationship between speed and distance. For example, the teacher might say, "Everyone seems to agree that this car rolled farther. Does anyone remember whether this was also the faster car?" At this point, the experiment may need to be repeated so that the children can verify their answers.

When children move to the elapsed-time component of the activity, an inverse relationship is encountered. The faster car takes less time, and therefore fewer claps, to reach the bottom of the ramp. This may be difficult for some children to grasp, so repeated experiences are desirable. The teacher might acknowledge this by saying, "Do you believe that! The fast car only got 3 claps, but the slow car got 6 claps. Should we try it again? Whose turn is it to roll the cars?" After several repetitions, children can begin to talk about why the faster car takes less time to reach the bottom, and therefore receives fewer claps.

Large-Group Activities

In order for children to develop measurement concepts, they must be the ones to do the measuring. For this reason, most of the measurement curriculum should unfold at the individual or small-group level. Whole-group experiences, however, are useful for introducing measurement concepts, such as by reading stories related to measurement, or by focusing attention on measurement problems that can then be explored in smaller groups. Large-group experiences also serve as opportunities to recap what has been learned and share ideas about measurement concepts.

Many of the activities discussed in the previous sections could start and conclude during group times. For example, "Inchworms" might begin with the teacher reading the Leo Lionni (1960) book to the class and measuring an object with the inchworm manipulative pieces. Children could then discuss things that they would like to measure in the classroom, carry out their measurements individually or in small groups, and report their measurements back to the large group. Interest may develop among the older children when different people come up with different measurements for the same object. This can lead to a discussion about how to make measurements more accurate.

An activity such as Container "Brainer" might also be introduced during group time. Teachers could chart children's estimates of which container they think will hold the most scoops of water. This may lead children to more focused play with the containers during choice time. As the activity is implemented, teachers can make note of children's discoveries and later share them with the larger group.

The two activities that follow are designed to be introduced at the large-group level, implemented individually or in small groups, and then revisited with the whole class.

ACTIVITY 6.8

How Tall Am I?

Materials

The following materials are needed for this activity:

- Salt boxes or cocoa containers (at least 10)
- Half-gallon cardboard milk cartons (at least 10)
- Tape measure
- Digital camera

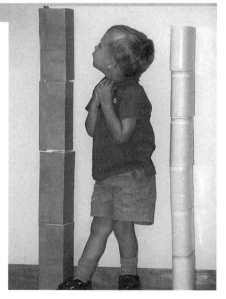

Description

This activity challenges children's thinking by providing two different sizes of containers to use as measuring units. The salt boxes are about $5\frac{1}{2}$ inches tall, and they stack easily. The half-gallon milk cartons, thoroughly washed and dried, form sturdy blocks when the open end of one is inserted into the open end of another to form a double-thickness box. The milk-carton blocks can be stacked vertically and are about $10\frac{1}{2}$ inches tall.

The activity can be introduced at group time, perhaps with a book such as *We Are All Alike: We Are All Different* (Cheltenham Elementary School Kindergartners, 2002). One of the ways that children are alike is that they are all growing. A way in which they are different is their current heights. After reading the book, teachers can demonstrate the result of measuring the same person with different-sized units. For example, Jamie might be as tall as eight salt boxes, but only four milk cartons. A measurement area can be set up in the classroom so that children can help one another with measurements. Teachers can help children record their heights and take a digital photograph of each child standing in between a column of salt boxes or similar containers and a column of milk cartons. The photographs can be enlarged and assembled into a class book to share at group time. Children can return again and again to these photographs to visualize why there are more salt boxes than milk cartons. If desired, teachers can also use a measuring tape to measure each child.

Math Discussions

Several important conversations are likely to emerge during this activity. First, there is the interesting dilemma of needing more salt boxes than milk cartons to measure the same person. Teachers should ask questions that focus children's attention on this relationship. They might say, "Are the two towers of blocks the same height as Ellen? Are there as many milk cartons as salt boxes? Why not?"

Children who do not yet have strong counting skills may disagree as to how many of each type of unit are used in the measurements. The teacher can mediate by asking children to show one another how they have arrived at their answers. This is a helpful application of the Reasoning and Proof process standard.

Because the blocks must be stacked, measurement errors, such as not having the units touch or not starting the measurement at the bottom of the object, typically should not be an issue. However, some children may notice that the blocks do not provide exact measures. For example, they may extend beyond the child's head or not reach quite far enough. In these cases, the teacher should ask for ideas. The children may be satisfied by saying that a child is *almost 9* salt boxes tall or that a child is *about* 10½ salt boxes tall. Some children may suggest a smaller unit of measure, particularly if they have had many experiences with measurement. One-inch cubes or the previously discussed inchworm measures could be suggested. This might result in a measurement of nine salt boxes and four cubes. This is also a good time to show the children a standard measure and discuss why it is divided into smaller units.

ACTIVITY 6.9

Is It Larger? Is It Smaller?

Materials

The following materials are needed for this activity:

- Copy of the children's book *Is It Larger? Is It Smaller?* (Hoban, 1997)
- Digital camera

Description

Measurement is rooted in the perceptual comparisons that children make between objects, accompanied by the relevant vocabulary. The Tana Hoban book focuses on these foundational concepts through clear photographs of objects that are identical except for their sizes. Teachers can start by sharing this book with their class and discussing each of the pictures. Throughout the week, children can look for objects in their classroom or school that are the same except for their size. A digital photo of the two objects placed next to one another highlights the size difference. These photos, along with each child's description, can form the pages of a class big book that can then be shared during group time.

Math Discussions

Some children may quickly find objects to compare; however, other children may need some guidance. In these cases, the teacher may suggest that a child accompany her on a "size hunt." The teacher might point out a large object, such as a cushion in the reading area, and then suggest that they look around for a smaller version. Other children may want to join in on these size hunts and share their observations.

As children look at the photographs in the class book, teachers can direct their attention to additional measurement comparisons. For example, a child might have a page that shows a big bucket and a little bucket. The teacher might ask whether one bucket is also taller or wider than the other bucket, and indicate with gestures what this means. In some cases, two objects may differ in one dimension, but not in another. A photograph of two pencils might show a long and a short pencil that have the same width. Children like to read books many times, particularly if they have participated in creating them. Each time the class size book is read, teachers can draw attention to a new measurement comparison.

Measurement Throughout the Curriculum

Given the many measurable attributes of objects, as well as the inquisitiveness of children, it makes sense to include a small measurement center in the classroom. Rather than having a balance available only once or twice a year for a particular activity, it could be ready for use whenever children have a measurement question involving mass. Tools for measuring length and volume could also be housed in the center. Interlocking manipulative units of various sizes, rulers, tape measures, pencils and paper, and measuring cups might be some of the tools housed in this area. Having a selection of measurement tools to choose among for a given problem encourages children to think about the attribute they wish to measure and the appropriate tool to meet this need. Understanding the range of measurement tools, along with the experience to select an appropriate tool for a given measurement situation, is a key component of the Measurement standard. The examples that follow illustrate potential measurement situations throughout the classroom.

Music Area

Many musical instruments come in different sizes: triangles, wood blocks, drums, and so forth. Larger instruments have lower tones than smaller instruments. When teachers make several sizes of the same instrument available for exploration, children can learn the relationship between size and pitch. A lovely instrument can be made by cutting aluminum or copper pipe to various lengths, sanding the edges, and laying the pipes across a frame made of packing foam, with grooves cut to hold the bars. Children like to seriate the pipes as well as listen to their beautiful tones.

Cooking

Cooking activities give children real-life experiences in measuring dry and liquid volumes. Teachers should draw attention to the relative sizes of measuring cups and spoons. An activity that arouses measurement excitement in children is peeling an apple with a hand-cranked peeler. The peel from what appears to be a small apple comes off in a *very* long string. Children find this amazing and often want to measure the peel.

Gardening

Planting seeds is a common activity in preschool and kindergarten. If several different seeds are planted by each child, the growth of each plant can be measured and charted. Children can measure their plants daily by marking their height on a straw or stick pushed, for that purpose, into the dirt next to each plant. Some children may want to use tools from the measurement center so that they can record the measurements in units.

Outside Area

Measuring tracks is an interesting outside activity. If children dip the soles of their shoes in a tray of water, the outline becomes visible when they step on the pavement. Children can use chalk to trace around their shoe prints and then measure them. Tricycles make an interesting series of tracks of three parallel lines if they are ridden through a wet spot on the concrete. The tracks become fainter the farther the tricycle gets from the water. Children may want to pace out the length of one another's tricycle tracks.

Dramatic Play

Dramatic play accessories do not always have to be the same size. Instead of the standard sets of play dishes, in which all of the cups and plates are the same size, teachers might consider substituting various sizes of plates, glasses, cups, and silverware. This gives children the opportunity to make size comparisons while they play. It also provides opportunities for size-related words to enter the conversation. For example, the teacher might ask for the small plate, but the large cup, because he is more thirsty than hungry.

Literacy

Children's books can be used to generate interest in measurement or to solidify vocabulary. In addition to those books already mentioned in this chapter, *Measuring Penny*, by Loreen Leedy (1997), shows many different ways for a young girl to measure her dog. *How Big Is a Foot*, by Rolf Myller (1962), is an amusing story that highlights the need for a standard measurement unit. *Length,* by Henry Pluckrose (1995), uses clear photographs to illustrate measurement concepts.

UNIVERSAL DESIGN FOR SUPPORTING MEASUREMENT

As with all learning, teachers should expect to see in their children a developmental range of understanding for measurement concepts. For this reason, open-ended activities that allow children to use their own thinking strategies, while also observing the strategies of their peers, are desirable. Children who have cognitive delays may continue to rely on direct measurement comparisons for a much longer time than their peers. Teachers should remember, though, that even typically developing preschool and kindergarten children become confused when one of two sticks judged to be the same length is moved to a new position. Full understanding of measurement takes a long time to develop for all children.

Measurement often relies on visual input, so children with visual disabilities are at a disadvantage. Teachers should remember to include tactile dimensions to measurement activities. If flat objects are being measured, such as tracings of the outlines of children, the lines should be retraced with puffy paint for children with visual difficulties. This allows them to feel the dimensions that they are measuring. The interlocking inchworms previously referred to in this chapter are a good tool for children who are visually impaired, because they can feel the inchworm on the back of each block. This makes it easier for such children to quantify the units. However, traditional measuring tools, such as yardsticks and tape measures, have flat representations of units. These should be written over with puffy paint to make them accessible by touch.

FOCUSING ON MATHEMATICAL PROCESSES

- *Problem Solving*—Children become interested in measurement when they want to communicate the size of something important. This might be the height of a tall building that they have constructed, the length of a train track, or their own height. They may use direct comparison, such as placing two objects next to each other; use an intermediary object, such as cutting a string that is as long as the object; or line up units, such as pieces of paper, to solve their measurement problem.

- *Reasoning and Proof*—When making measurement comparisons, young children primarily use direct comparison. They may set two objects next to one another to prove that one is bigger than the other. Children who are using units to measure, such as older kindergartners, may count the units to prove that one container holds more than another, or that one object is longer than another.

- *Communication*—Children first communicate measurement concepts through comparison terms, such as "big" and "little," or "fast" and "slow." Preschool and kindergarten children may also compare an object to a remembered image. For example, they might say, "That dog is a lot bigger than my dog." They also make measurement comparisons based on function, such as, "That horse is too big to fit in this truck."

- *Connections*—Measurement is directly connected to Number and Operations, and to Geometry. As children develop the concept of a measurement unit, they must quantify and compare the units. In some cases, such as when they build a tower, they may even add units. Measurement is also closely connected to Geometry, in terms of both the definitions of shapes and the location of objects.

- *Representation*—Young children often represent measurement relationships with words and gestures; for example, they may hold their hands far apart to indicate that something is big. They may also represent size through their drawings, with parents shown as larger than children. These representations, however, are far from exact. A person in a child's drawing may be nearly as large as a house.

ASSESSING MEASUREMENT

The same methods used to assess learning in Geometry—work samples, photographs, anecdotal notes, and teacher-designed checklists—are useful for documenting understanding and application of measurement concepts. Of course, the more often that measurement is incorporated into the curriculum, the more opportunities are created for learning and assessment. Measurement is often underrepresented in preschool and kindergarten lesson plans.

Work samples are generated when activities include fill-in sheets, such as the "Inchworm" activity presented in this chapter. Also, children sometimes incorporate measurement into their artwork, such as creating paint imprints with two sizes of cookie cutters, perhaps arranged in a pattern. Individual pages from class books can be inserted into children's portfolios for documentation, once the books are no longer needed.

Photographs can quickly capture children's use of measurement by direct comparison. Children may set a toy animal inside a building that they have constructed to see whether it will fit, or carefully roll a truck up to a bridge to see whether it will fit underneath. Photographs can also preserve more elaborate uses of measurement. For example, children may align manipulative units next to an object that they are measuring or even attempt to construct their own measuring tape. In these instances, one photograph can take the place of many words.

Anecdotal notes are invaluable for preserving children's thinking because teachers can notate exactly what children say and do. Teachers will want to record the measurement strategy used by the child, along with any misconceptions, such as failure to start the measurement at the end point of the object, using measurement units that are not all the same size, or placing measurement units with gaps in between them.

Checklists are a quick way to capture data on many different children, especially when a measurement activity has been planned. Teachers might want to create columns, one each for direct comparison, use of an intermediary object, and unitizing. A space for comments is important because not every piece of important information can be captured on a checklist.

SUMMARY

The Measurement standard includes two main components: 1) understanding of the measurable attributes of objects and 2) using appropriate tools and techniques. By the end of second grade, students are expected to compare and order objects according to attributes such as length, volume, and weight, and to select appropriate nonstandard and standard units and tools. Students are also expected to use repetitions or multiple copies of a unit, and to develop mental images of common measures to use as reference points.

Preschool and kindergarten children have several strategies for judging relative size. They use perceptual comparisons when both of two compared objects are present and can use the memory of an object as a norm for comparison with other objects. Young children also use functionality to make size decisions.

The development of measurement concepts begins in infancy but continues to evolve over many years. Children begin by using words such as "big" and "little" to represent measurements. Next, they directly compare two objects to determine whether one is larger than the other. The idea of a unit of measure that can be counted to produce a measured result takes a long time to develop. At first, children usually select nonstandard units, such as blocks or fingers, to measure length. Measurement errors are typical. Children do not understand the need for contiguous units and may leave gaps in between their measuring units. They may not start measuring at the end of the object and may arrange the units in a wavy rather than a straight line. Despite their limitations in reasoning, young children do successfully invent strategies to solve measurement problems.

The measurement curriculum should center around small-group experiences, because children need to conduct the measurements themselves in order to understand the underlying concepts. The conversation that emerges when peers work together can further children's understanding of measurement. Many activities work best when introduced to the whole group, implemented at the small-group level, and then revisited and discussed with the whole class. Although measurement activities may evolve naturally from the play of young children, they are more likely to occur when teachers intentionally plan activities that encourage size comparisons and unitizing. Teachers should be prepared to offer comments, ask questions, or model measurement concepts through play.

ON YOUR OWN

- Create a list of children's books related to measurement. The local library is a good resource.
- Think of five different ways to incorporate measurement concepts into transitional activities with young children.
- Plan a class project that would incorporate measurement concepts.

REFERENCES

Andrews, A.G., & Trafton, P.R. (2002). *Little kids—Powerful problem solvers: Math stories from a kindergarten classroom.* Portsmouth, NH: Heinemann.

Bausano, M.K., & Jeffrey, W.E. (1975). Dimensional salience and judgments of bigness by three-year-old children. *Child Development, 46,* 988–991.

Brannon, E.M. (2002). The development of ordinal numerical knowledge in infancy. *Cognition, 83,* 223–240.

Cheltenham Elementary School Kindergartners. (2002). *We are all alike. . . We are all different.* New York: Scholastic.

Fabricius, W.V., & Wellman, H.M. (1993). Two roads diverged: Young children's ability to judge distances. *Child Development, 64,* 399–414.

Gelman, S.A., & Ebeling, K.S. (1989). Children's use of non-egocentric standards in judgments of functional size. *Child Development, 60,* 920–932.

Hoban, T. (1997). *Is it larger? Is it smaller?* New York: Greenwillow Books.

Inhelder, B., Sinclair, H., & Bovet, M. (1974). *Learning and the development of cognition.* Cambridge, MA: Harvard University Press.

Johnson, G. (2008). *The ten most beautiful experiments.* New York: Knopf.

Kamii, C. (2000). *Young children reinvent arithmetic: Implications of Piaget's theory* (2nd ed.). New York: Teachers College Press.

Leedy, L. (1997). *Measuring Penny.* New York: Henry Holt and Company.

Lionni, L. (1960). *Inch by inch.* New York: Astor-Honor.

Myller, R. (1962). *How big is a foot?* New York: Random House.

Miller, K.F., & Baillargeon, R. (1990). Length and distance: Do preschoolers think that occlusion brings things together? *Developmental Psychology, 26,* 103–114.

National Council of Teachers of Mathematics. (2006). *Curriculum focal points.* Reston, VA: Author.

National Council of Teachers of Mathematics. (2000). *Curriculum and evaluation standards for school mathematics.* Reston, VA: Author.

Piaget, J. (1952). *The child's conception of number.* New York: Norton.

Piaget, J., & Inhelder, B. (1967). *The child's conception of space.* New York: Norton.

Piaget, J., & Inhelder, B. (1969a). *The psychology of the child.* New York: Basic Books.

Piaget, J., & Inhelder, B. (1969b). *The mechanisms of perception.* New York: Basic Books.

Piaget, J., Inhelder, B., & Szeminska, A. (1960). *The child's conception of geometry.* London: Routledge.

Pluckrose, H. (1995). *Length.* Chicago: Children's Press.

Raven, K.E., & Gelman, S.A. (1984). Rule usage in children's understanding of "big" and "little." *Child Development, 55,* 2141–2150.

Reggio Children. (1997). *Shoe and meter.* Reggio Emilia, Italy: Author.

Sena, R., & Smith, L.B. (1990). New evidence on the development of the word big. *Child Development, 61,* 1034–1052.

Wadsworth, B. (1989). *Piaget's theory of cognitive and affective development* (4th ed.). White Plains, NY: Longman.

Wolf, Y. (1995). Estimation of Euclidian quantity by 5- and 6-year-old children: Facilitating a multiplication rule. *Journal of Experimental Child Psychology, 59,* 49–75.

Yuzawa, M., Bart, W.M., Kinne, L.J., Sukemune, S., & Kataoa, M. (1999). The effects of "origami" practice on size comparison strategy among young Japanese and American children. *Journal of Research in Childhood Education, 13*(2), 133–143.

7

Developing Concepts About Data Analysis and Probability

Put the names of all the people who voted for green apples on a green piece of paper. Put the names of all the people who voted for yellow apples on a yellow piece of paper. Put the names of all the people who voted for red apples on a red piece of paper. Then we can tell which apple was the most popular.

—Alice, astute preschool child

Children begin analyzing physical data as part of their earliest explorations. As they interact with their environment, they take in raw data, or information, through their senses. Through the mental sorting of this data, children learn to adapt to their environment. By the age of 3–4 months, infants search for a milk-producing nipple when hungry and reject other objects placed in their mouths, indicating that mental categories of milk-producing and non-milk-producing have already been formed (Wadsworth, 1989). Toddlers may have already discovered that the family dog tolerates having its tail pulled, but the cat scratches. Therefore, they have also learned that animals resembling cats have to be approached with caution.

As children develop, they appear to organize data purely for the mental enjoyment or stimulation. Young preschoolers may create rows of pegs grouped by color even though no one has suggested this idea to them. They may arrange "parades" of toy animals, grouped by type of animal and organized from largest to smallest within those groups. Toys and other objects in their environment become their data, and they group, rearrange, compare, and quantify as they explore these materials.

Older preschool and kindergarten children, who are increasingly aware of the ideas and opinions of their friends, become interested in voting to determine outcomes of class events, such as what type of fruit to have for lunch, or to compare opinions on various subjects, such as favorite zoo animal. Teachers may tally these votes on class bar graphs so that children can more easily quantify and compare them.

Young children also make predictions. When Mike's dad arrives at school with birthday cupcakes, children may assume that there will be one for every child because this is what has happened before when parents brought in treats. However, many of the predictions that children make are remarkably flawed due to lack of experience and developmental levels of reasoning. For example, one preschool class that was involved in a tree project made daily predictions about what colors they thought the leaves on the playground trees would be on the following day. Even though the leaves remained green day after day, color predictions such as pink and blue were common, perhaps because these were colors that children wanted the leaves to be.

THE DATA ANALYSIS AND PROBABILITY STANDARD

The Data Analysis and Probability standard is envisioned by the National Council of Teachers of Mathematics (NCTM) as an important component of mathematics education that should span all of the grades, from preschool through Grade 12 (NCTM, 2000). Understanding of content within this standard enables students to formulate questions and collect and analyze data to answer them. These data can come from the student's world—sports or interesting current events for older students, and topics related to families, food, pets, or school projects for younger students. Although data analysis is not considered one of the three curriculum focal points for preschool and kindergarten children, it is viewed as closely related to geometry and measurement because its focus is also on the attributes of objects (NCTM, 2006). In addition, the rules that are created for sorting and classifying objects connect data analysis to algebra, and the quantification of data ties it to number concepts. The Data Analysis and Probability standard, therefore, is closely connected to all of the other mathematics standards.

The Data Analysis and Probability standard has four main areas of focus:

- Management of data, which includes formulating data-related questions as well as collecting, sorting, and displaying data

- Use of appropriate methods to analyze data

- Predictive use of data

- Beginning concepts about probability

Data-related questions for preschool and kindergarten children come directly from their environment and experiences. How many children are buying lunch versus packing? What games had the most votes for field day? How many children have birthdays this month? What flavor of ice cream did the class decide to make? What name was selected for the new hermit crab? These are the types of questions that are important to young children and, therefore, focus their attention on collecting and analyzing relevant data.

An important source of data for young children comes from voting. Kamii (1982, 2000) recommends voting as a regular activity in preschool and kindergarten classrooms because it encourages children's autonomy while also providing numerical quantities to compare. Voting by show of hands, however, can be problematic for young children, who often want to vote for every choice (Moomaw & Hieronymus, 1995). Allowing children to vote one at a time eliminates this problem. By writing children's names on slips of paper to record their votes, and then displaying the votes on a graph, teachers create a semipermanent record of the votes that children can refer to as they analyze the data. The graphic display enables children at the global, one-to-one correspondence, and counting stages of quantification to make comparisons among the categories. It also provides a model for organizing and displaying data.

Children also organize data by sorting and classifying groups of objects. In Chapter 2, divided trays or small, separate containers were suggested as ways to encourage children to place objects into groups. When data have been sorted and are ready to analyze, graph paper or cardboard strips divided into boxes can help children make comparisons. The grids encourage children to place one object in each box, in a one-to-one correspondence fashion. This makes comparison of the various quantities more accurate because children can align the rows, rather than simply look at piles of objects and make a global estimation of which pile has more.

Use of data to form inferences and make predictions is not one of the expectations for the preschool to Grade 2 grade band; however, this should not discourage teachers from encouraging children to talk informally about these concepts. In the Ramp Races activity described in Chapter 6, children raced toy cars down inclines with different slopes. In activities such as this, children quickly determine that the car on the steeper slope goes faster. Teachers who ask for a turn are likely to find themselves relegated to the slope that is less steep, thereby assuring that they will always lose. Children clearly have made an inference based on their experiences with the ramps and can now accurately predict which car will be likely to win the race. For this reason, physical knowledge activities, which provide children with an immediately observable reaction, are good choices for initial experiences in making predictions.

Similar to prediction and inference, probability is treated informally in the preschool and kindergarten grades. Preschool children are still struggling with reality versus fantasy, so the concepts *possible* and *impossible* can be challenging. Adults may have difficulty convincing a child who has seen Superman fly on the movie screen that people do not fly. Nevertheless, the conversations that adults have with children help them begin to sort out what is likely or unlikely to happen. For example, in a kindergarten class of mine, children who had a rabbit as a class pet accurately inferred that pet rabbits, in general, live in hutches and eat rabbit pellets. However, these same children believed that wild rabbits lived in little houses and wore jackets with gold buttons!

THE DEVELOPMENT OF CONCEPTS ABOUT DATA ANALYSIS AND PROBABILITY

Data Analysis

Although there is little research to date regarding young children's development of data analysis concepts (Clements & Sarama, 2007), preschool and kindergarten teachers who have used graphing as a component of their mathematics curriculum provide anecdotal information. Baratta-Lornton (1976) described a sequence of graphing activities that she developed for her kindergarten classes to increase their abilities to organize information, make comparisons among groups, and problem solve. For their first experiences with graphs, children used real objects and compared two, and then three, groups. Next, they used pictures of objects to compare two and three groups. This was followed by graphing with real objects to compare four groups, followed by picture graphs comparing four groups. Finally, children moved to symbolic graphs, in which a symbol such as an "X" could stand for a cookie. Baratta-Lornton believed that this progression was important because it moved the children from concrete to more abstract representations.

Other educators have developed graphing as a way to visually represent children's voting responses and to increase opportunities for children to make numerical comparisons (Moomaw & Hieronymus, 1995). These group graphing experiences were designed primarily for older preschool and kindergarten children who had already had many experiences comparing groups of objects during their play. The children's names were recorded on voting tags, which were then placed on bar graphs. Moomaw and Hieronymus reported that the use of specific names helped children conceptualize a one-to-one correspondence relationship between each child's vote and its representation on the graph. Children did not seem to need real objects or pictures to understand where each vote should be placed. Most important, use of the children's names preserved their votes. When it was time to go home, children led their parents to the graphs to discuss who had voted for what and to interpret the results. Long after the initial voting had taken place, children would return to the graphs to count votes, make comparisons, and discuss the results.

Just because young children are able to interpret a graph that is constructed as part of a group experience does not mean that these children could create such a graph on their own. To learn more about young children's graphing concepts, the author designed an individual graphing activity for a summer prekindergarten class. All of the children had participated in group graphing activities as preschoolers and would be entering kindergarten in the fall. Each child was given a bag containing from one to five goldfish crackers, one to five vanilla teddy bear grahams, and one to five chocolate teddy bear grahams, along with a sheet of 1-inch graph paper that had a drawing of each type of cracker along the bottom row. The children were asked to graph the crackers in their bag prior to eating them so that they could remember and compare what was in each person's bag. Children created six different types of graphic representations for their crackers. Although all of the children showed some conceptual knowledge of graphing, varying levels of understanding were apparent. Figures 7.1–7.6 are representative of the types of graphs that the children created.

Some of the children started their graphs by lining up the crackers, one per box, in columns above their respective drawings. Philip, whose graph appears in Figure 7.1, then drew a vertical line through all of the boxes that held a cracker; however, for the column that contained five chocolate bears, he extended his line all the way to the top of the paper. Philip showed an understanding of the importance of one-to-one correspondence in the placement of crackers on his graph. As he formed columns of various types of crackers, he started at the bottom and did not skip any spaces. Philip also realized that his lines could represent the number

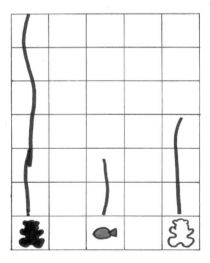

Figure 7.1.   Philip's graph. (From Moomaw, S., & Hieronymus, B. [1995]. *More than counting* [p. 193]. St. Paul, MN: Redleaf Press; reprinted by permission.)

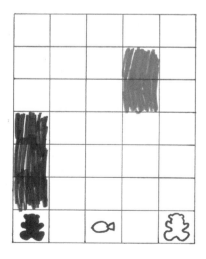

Figure 7.2.   Kenan's graph. (From Moomaw, S., & Hieronymus, B. [1995]. *More than counting* [p. 193]. St. Paul, MN: Redleaf Press; reprinted by permission.)

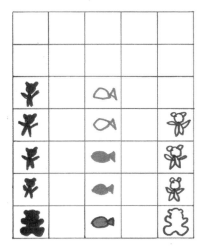

Figure 7.3. Anthony's graph. (From Moomaw, S., & Hieronymus, B. [1995]. *More than counting* [p. 193]. St. Paul, MN: Redleaf Press; reprinted by permission.)

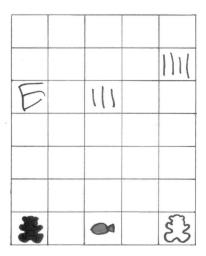

Figure 7.4. Douglas's graph. (From Moomaw, S., & Hieronymus, B. [1995]. *More than counting* [p. 193]. St. Paul, MN: Redleaf Press; reprinted by permission.)

of each type of cracker, and he ate all of his crackers once he had completed his graph. His only error was to extend to the top of the page the line meant to represent five crackers. Perhaps this seemed like a lot of crackers to him, and he therefore showed a global representation.

Some children placed their crackers inside the boxes on the graph paper, but did not consistently align them in columns. Kenan's graph, which appears in Figure 7.2, is an example of this type of representation. Kenan used markers that matched the colors of his chocolate bears and goldfish crackers to color a corresponding number of boxes on his graph. He did not represent his vanilla teddy grahams, perhaps because they were the color of the graph paper. Although Kenan showed a one-to-one correspondence relationship for two of his types of crackers, he did not yet understand the need to begin the columns at the bottom of the graph paper in order to compare the heights of the columns. He also did not place his crackers in columns directly above the pictures that represented them.

Some children correctly graphed their crackers in the appropriate columns, but felt the need to draw a picture of each cracker. Anthony's graph in Figure 7.3 is representative of this type of thinking. Rather than simply coloring in boxes in the appropriate columns, or drawing a line through the boxes, as Philip did, Anthony drew a symbolic representation of each cracker. From Anthony's perspective, one picture at the bottom of a column was apparently not sufficient to represent all of the crackers in that column.

Some children used symbols to represent the totals on their graphs. Recall that, from the Piagetian perspective, symbols, unlike signs, resemble what they represent and can be created by the child (Kamii, 2000). In Douglas's graph (Figure 7.4), hash marks at the top of each column represent the number of crackers that were originally placed there.

Some children used both symbols and signs to represent their crackers. Heidi's graph (Figure 7.5) included a picture of each cracker, aligned in the appropriate column, along with a numeral to summarize the totals.

Only one child in the class seemed to realize that she could represent her crackers by simply coloring in an appropriate number of boxes in the column that represented that cracker. Ping used only one color of marker for the entire graph (Figure 7.6). She explained that she did not need to change colors or draw pictures. She could tell how many of each type of cracker she had by simply looking at the columns.

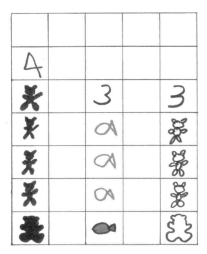

Figure 7.5. Heidi's graph. (From Moomaw, S., & Hieronymus, B. [1995]. *More than counting* [p. 194]. St. Paul, MN: Redleaf Press; reprinted by permission.)

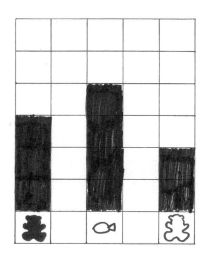

Figure 7.6. Ping's graph. (From Moomaw, S., & Hieronymus, B. [1995]. *More than counting* [p. 194]. St. Paul, MN: Redleaf Press; reprinted by permission.)

The work of these students shows that, through their group graphing experiences, they constructed fundamental concepts related to the graphic representation of data. This should encourage preschool and kindergarten teachers to explore graphing with groups of children in their classrooms. When children have had many experiences with graphing, they may begin to collect their own data to answer questions that they have formulated. In one preschool classroom, the teacher was surprised one day by three girls who were walking around the room with a clipboard. The girls explained that they wanted to find out each child's favorite color. They interviewed their friends and, using phonetic spelling, noted their color preferences. They asked the teacher to help them graph the results during group time. This anecdote shows that at least some preschool children understand the concept of a graph to display and analyze data in order to answer a mathematical question.

Probability

Piaget describes children's understanding of probability as developing in three stages (Piaget & Inhelder, 1976). During the first stage, which includes most children younger than the age of 7, children are not able to distinguish between events that are necessary versus those that are merely possible. This is not surprising, because Piaget documents children at this age as being unable to reverse their thinking about events; therefore, they cannot form logical deductions (Piaget, 1965). For example, if a row of objects is spread apart, children cannot mentally picture the reverse operation of pushing the objects back together; therefore, it is perfectly plausible to them that there are now more objects in the row. Piaget reports no evidence of a concept of uncertainty during this first stage (Shaughnessy, 1992). Thus, if children see instances of event A appearing more frequently than event B, they predict that event B will happen because it has been skipped too often. Finally, according to Piaget, children at this stage do not perceive probability as a ratio. If they are shown two boxes of marbles, one with three black marbles and one white marble, and the other with six black marbles and two white marbles, children often decide that they are more likely to pick a black marble from the box with six black marbles, even though the probability of picking a black marble from either box is the same. More recent research largely confirms Piaget's descriptions of children's thinking (Jones, Langrall, & Mooney, 2007; Shaughnessy, 1992). For example, Metz (1998) reports that few kindergarten children are able to interpret random phenomena. If colored marbles are placed in a box that is then tilted, they expect the marbles to return to their original position when the box is set level again.

Estimation

Although *estimation* and *probability* are concepts that differ somewhat from each other, it can be argued that the two are related, particularly in the minds of young children. Estimation involves a logical approximation of a result when all of the data are not given or are uncertain. Rather than having the children merely guess, the goal is for children to use the information that they know to make a logical prediction. According to Bayesian theory, probability is a way to represent the degree to which a person, given the evidence, can agree with a statement. If a child estimates that there are seven buckeyes in a clear jar, the child is expressing a likelihood that there are about seven buckeyes in the jar, even though the child cannot see (because some marbles are in front of others) and, therefore, count all of them. Because of the relationship between probability and estimation, the subject of estimation is included in this section.

Research indicates that some developmentally advanced 3-year-old children are able to invent decomposition strategies to estimate collections of three to six items (Baroody, Lai, & Mix, 2006). In the same study, 4-year-old children who were allowed to briefly view collections of three to five items were able to estimate their size to within about one point of their actual value; however, estimates were less accurate for collections of 7–11 items. Some of the children attempted to count the items that were visible. Of these children, some gave the last number they counted as their estimate, whereas others gave a number that was higher than the last number counted. This research shows that preschool children have some idea of logical possibilities for the cardinal value of small groups of objects. The authors of the study suggest that developmentally appropriate estimation activities can begin during the preschool years, with quantities just beyond the children's level of competence. Children should be encouraged to communicate their estimation strategies.

DESIGNING THE DATA ANALYSIS AND PROBABILITY CURRICULUM

Curriculum designed for the Data Analysis and Probability standard should encompass managing data, analyzing data, making predictions based on those analyses, and exploring beginning concepts of probability. The primary method that young children use to manage data is to sort them into groups. Once this has been accomplished, children may make comparisons among the groups. For example, a child playing with toy animals may group all of the same type of animal together. The child may then notice that there are a lot of sheep, but only one chicken. In a similar fashion, children playing in the dramatic play area may group toys on the basis of function, such as by placing one cup on each saucer. If one saucer is missing a cup, this is quickly noticed. Although children may spontaneously sort and group objects, this important concept should be included regularly as a planned part of the mathematics curriculum. Teachers can assemble collections of objects for children to sort and classify according to various attributes. The topic of sorting and classifying, which has already been discussed in Chapter 4, is extended in this chapter to include individual and small-group graphing activities. Teachers should remember that goals in both Algebra, and Data Analysis and Probability, are encompassed in sorting and classifying activities.

Graphing data, including responses from voting, should be a main component of the data analysis curriculum in preschool and kindergarten. As explained earlier, it is helpful to give young children a forced choice when voting; otherwise, children often want to vote for everything. For example, rather than asking who wants to name the new goldfish Spotty, the teacher might ask, "Zina, do you want to name the goldfish Spotty, Goldilocks, or Fishy?" When each child's vote is then recorded with a name tag on the graph, children draw a one-to-once association between their vote and its concrete representation on the graph.

A problem that can develop when teachers graph children's votes is the desire of many children to be on "the winning team." As soon as a few votes appear in a particular column, all of the children may begin voting for that item. In addition to creating data that are less interesting to analyze, this type of response is antithetical to autonomy. A way around this problem is for teachers to collect children's responses prior to group time. The results of these secret ballots can then be graphed with the whole group. It is important to print the children's names on their voting tags ahead of time. Children lose attention if the graphing process takes too long.

Teachers can also introduce beginning probability concepts into their curriculum planning. For example, on a sunny day in May, teachers might ask the class whether it is likely or unlikely to rain or snow during outside time. When reading stories, teachers can interject probability questions. Do the children think that it is likely or unlikely that the chicken will be eaten by the fox? Is it possible or impossible for a teddy bear to walk and talk? It is important to also ask children to justify their responses.

Teachers can also plan estimation activities to help children focus on logical deductions. Of course, this type of reasoning takes time to develop. As noted earlier, estimation should involve relatively small quantities that are at the level of the child's competency or just slightly above. Activities that support data analysis, making predictions, and probability concepts are included in the sections that follow.

Math Talk

A critical component of data analysis is the conversation that takes place after the data have been organized. Teachers' questions and comments guide children to consider categories that have the most and fewest representations, categories that may have received the same number of responses, and differences among the categories. On the basis of these discussions, teachers can ask children to make predictions about future data. The following are examples of teacher questions and comments that can guide children in interpreting graphic representations of data:

- *Which column on our graph has the most votes? How can you tell?*

- *Which column has the fewest votes? Did any animal on the graph receive less than one vote?*

- *Can you tell from the graph how many people have brought in field-trip permission forms? How many people still need to bring theirs in?*

- *It looks as though a lot of children want to have fish sticks for lunch. Let's count the votes. We have 10 children who want fish sticks, 2 who want macaroni, and 3 who want pizza. According to these votes, what do you think Joel will vote for when he comes to school tomorrow? Can his vote change the winner of the lunch vote?*

- *A lot of people wore flip flops today, and a lot of people wore Velcro shoes. How many more people wore flip flops than Velcro?*

- *Brian says that the elephant is ahead by one vote over the giraffe for the animal we should visit first on our zoo trip. What will happen if Annie votes for the giraffe when she gets back to school?*

Teachers can also help children decide what data they need to collect to answer specific questions. Suppose that a class is planning a family dinner for the end of the year and the children are going to make a large salad. The teacher might suggest that the children ask their parents what their favorite salad vegetable is. The responses can be added to a class graph, and on the basis of the information, the children can help the teacher decide how many of each type of vegetable to buy.

Making predictions is an important part of scientific inquiry. When children are involved in scientific observations and explorations, teachers can ask them to predict what will happen. Formulating and checking a prediction helps focus children's attention on outcomes. As children collect data, they can dictate or record their observations in an individual, small-group, or class science log. Periodically, the data can be reviewed and compared with the children's predictions.

Finally, teachers can become more intentional about adding the language of probability to their conversations with children. *Likely, unlikely, possible, impossible, probably, never, always, might,* and *maybe* are terms that children can begin to understand if they are used in a relevant context. Some examples follow.

- *Look at all those dark clouds. It's* probably *going to rain soon.*

- *The sun is coming out. We* might *be able to go outside after all.*

- *I have* never *heard our pet rabbit talk. Have you?*

- *The gerbils* always *tear up the paper tube as soon as we put one in their cage.*

- *Are you more* likely *to hear the maraca with beans in it or the maraca with cotton balls in it?*

- *This book says it's* necessary *to let the goldfish's water stand for a day before we put the fish into it. Otherwise, the fish will not be able to breathe.*

- *Now that we've cooked our eggs, is it* possible *or* impossible *to put them back the way they were before?*

Individual or Small-Group Activities

There are several ways to incorporate data analysis and probability concepts into daily individual or small-group explorations. First, collections for sorting and classifying can be extended to provide graphing opportunities. This encourages children to compare sets, which also incorporates number sense. Second, activities that encourage children to make estimates of quantity can be available for use throughout the week and then examined as a group during circle time. Finally, prediction logs related to physical-knowledge activities encourage children to form hypotheses, collect data, and then compare the results to their predictions. The activities in this section are examples that teachers can use as models to create their own curriculum activities according to the interests of their class.

Graphing Collections

Chapter 4 offers suggestions for collections of objects that teachers can assemble to encourage children to sort and classify on the basis of various attributes. When children have become accustomed to playing with collections, grouping and regrouping the items in different ways, teachers can extend the activity to encourage organization of the data and comparison among the groups. A sturdy piece of cardboard, foam board, or light-weight wood, divided into columns and rows, provides a framework for individual or small-group graphing. Children may start by simply placing one item in each box. Teachers' questions and comments can guide children to the realization that, in order to accurately compare rows, they must start them at the same point and not leave any gaps. This reinforces important measurement concepts such as those discussed in Chapter 6. The graphing frame should be constructed so that it can be used repeatedly for a wide variety of collections.

ACTIVITY 7.1

Graphing a Dinosaur Collection

Materials

The following materials are needed for this activity:

- Collection of small dinosaurs that vary by color, size, and type of dinosaur
- Graphing board, made by dividing a piece of foam board, 20 × 12 inches, into 2-inch boxes outlined with permanent marker
- 2-inch cards, to label categories
- Children's book about dinosaurs (optional, but recommended)

Description

Many preschool and kindergarten children are fascinated with dinosaurs; therefore, they are eager to play with the collection. The activity encourages divergent thinking because the dinosaurs can be arranged in many different ways depending on an attribute selected by the children: color, type of dinosaur, size, plant-eater versus meat-eater, and so forth. Once the children have become confident in grouping the dinosaurs, the teacher can introduce the graphing board and suggest that the children use it to compare their sets. Children can use the cards to label their categories, or teachers can write the labels for them. The cards should be placed at the bottom of the board and attached with magnetic tape or Velcro. This makes it easy to change the categories when the items are regrouped. Some children may wish to report the results of their investigations at group time. Others may want to copy the large graph onto 1-inch graph paper that they can keep.

Math Discussions

Teacher scaffolding is an important part of this activity. Initially, children may need help deciding on categories for grouping the dinosaurs. Teachers can make suggestions, such as, "I notice that there are a lot of stegosauruses, the dinosaurs with the spikes on their backs. Should we put them together? Now you pick a type of dinosaur, and I'll help you find them." As always, teachers should encourage children to discuss why a particular dinosaur does or does not belong in a given group.

After the dinosaurs have been sorted, teachers can help children create the labels for the graph. As children begin transferring the dinosaurs onto the graph, disagreements may erupt. A dinosaur may be placed in the wrong column, or dinosaurs may be set randomly on the board. Children should be encouraged to voice their ideas. For example, the teacher might say, "Alex, why do you think that this dinosaur is in the wrong place? Do you agree, Steve?" The teacher's observations are also important, but they should be phrased in a nonjudgmental tone. For example, if a group of children places the dinosaurs randomly on the board, the teacher might comment, "Now I can't tell which dinosaurs you thought should go together. Can we put the same color of dinosaurs in a line so I can remember?" The subtlety in the teacher's language is important. She has not suggested that the children did anything wrong, but rather that *she* now has trouble remembering the groups. Contrast this with another teacher's comments: "Now you've mixed up all the groups. You have to put them in order on the graph." Although this teacher may be equally well meaning, the message conveyed to the children is that there is only one way to complete the graph, and the teacher knows what it is. Therefore, the children should look to the teacher for directions and copy her. This is not the message that teachers should send if they want children to develop into independent mathematical thinkers. Instead, teachers should accept children's mathematical representations as valid on the basis their current level of understanding, and phrase their comments to reflect this.

Teachers should be prepared to guide children when typical responses related to developmental understanding occur. For example, children may not start their rows at the same point on the graph, which may confuse them when they try to decide which category has the most items. The teacher might say, "Hmm, that's interesting, but it looks as if this row skips over some boxes. How would it look if you started this row next to the other one?" Remember that children who do not yet conserve discrete quantities, who will be most preschoolers and many kindergarteners, may believe that the number of items in the rows changes when the items are repositioned. It is through graphing activities such as this that children gradually revise their thinking.

Estimation

Estimation requires children to consider logical approximations when all of the data are not evident. Because young children are in the process of developing a sense of both number and logical relationships, it should not be surprising when the estimates of some children, particularly younger preschoolers, seem illogical. In classroom observations, when shown a jar of nine buckeyes, some preschool children gave an estimate of one, even though it was obvious that there were multiple buckeyes in the jar (Moomaw & Hieronymus, 1995). Other children gave their age as an estimate, an egocentric response indicating that, because the child's age is such an important number, it must be the number of buckeyes in the jar. Some children estimated the biggest number that they could think of, which varied from 20 to a "ka-zillion." With experience, however, children become much more accurate in estimating small quantities.

Estimation activities can be designed for children to explore during choice time in the classroom. When the activity is available over several days, children can return multiple times and reconsider their initial responses. They can also confer with their friends about their estimates. Later, the activity can be finalized during group time, and the objects actually quantified. Through comparing their estimates with the actual amounts, children begin to revise their thinking about quantity. During the estimation process, children must have some access to the data. When items are placed in a clear container, children may see some, but not all, of the items at any given time; therefore, they make a prediction according to what they can see. This is different from placing objects in an opaque container. If children cannot see any of the objects, they are merely guessing, not reasoning.

ACTIVITY 7.2

Estimating Shoes

Materials

The following materials are needed
for this activity:

■ Six to ten toddler shoes, placed in a
clear wide-mouth plastic jar

■ Basket of note cards, for children to
write their estimates, or estimation charts

■ Container for the completed estimates

Description

This activity can be placed on a bench, shelf top, or small table so that it is readily available to children. Teachers can vary the number of shoes in the jar on the basis of the
number-sense development of their class. For 3-year-olds, five or six shoes might be an
appropriate number, whereas for older kindergarten children, a number in the teens
might be in line with their thinking levels. In general, 6 to 10 items is a good number for
4- and 5-year-olds. Children may need to be reminded to include their name along with
their estimate on the note cards or chart. Although some teachers may choose to write
the children's names and number estimates for them, encouraging children to write a
numerical representation themselves provides ongoing assessment information about
how each child represents number. The results of the estimation activity can be revealed
and discussed during group time. All estimates should be treated as valid representations of the child's current thinking.

Math Discussions

Conversations that occur while children are making estimates are important because they
increase children's thinking. For example, if a child estimates that there is only one shoe in
the container, another child may reply that there are obviously more than that. Children
should be asked to explain and compare their thinking. The teacher might say, "Lisa, why
do you think that there are 10 shoes? Todd thinks that there are only four." Teachers may
want to list children's estimates on a chart prior to the grand opening of the jar at group
time. After counting the shoes, children can refer to the chart to see whether anyone's estimate was correct. They can also discuss predictions that were close to the exact amount.
This type of discussion may help children construct or solidify a mental number line.

Prediction and Inference

Mathematics is often used to quantify predictions made in other areas. Weather forecasters
quantify the probability of a particular kind of weather, and business administrators analyze
data to make predictions about future sales. For young children, math and science form a
natural connection, because children are constantly experimenting with materials to see
how they react. Quantifying the results of these explorations increases children's understanding of both mathematics and science. How many blocks did the pendulum knock
down? Will it knock down that many again? How many of the geometric solids roll down
the ramp? Do all of the round objects roll? Formally making predictions and then forming
inferences on the basis of how materials respond is an important part of scientific reasoning
that encourages children to think about probability.

ACTIVITY 7.3

How Many
Sank the Boat?

Materials

The following materials are needed
for this activity:

▪ Plastic dish pan, partially filled with
water

▪ Small plastic boat or hollow plastic
container

▪ Small plastic animals, with enough collective weight to cause the boat to sink

▪ Copy of the book *Who Sank the Boat?* by Pamela Allen (1996)

Description

This activity can be designed so that *almost* the same number of animals sinks the boat
each time. Weight differences among the animals and placement within the boat (or container) can cause some variation. This encourages children to think about probability. With
repeated attempts to sink the boat, children can discover *about* how many animals are
necessary. An observation log, on which children record how many animals were needed
to sink the boat on each trial, can help children judge how many to estimate on subsequent attempts.

Math Discussions

Math discussions for this activity should include recording data, making predictions based
on that data, and creating inferences. When children first attempt this activity, they are naturally drawn to the scientific aspects. Conversations should reflect this interest. Each time
an animal is added to the boat, the boat loses buoyancy and floats lower in the water.
Eventually, as more weight is added, an edge of the boat tips below the water line, and the
boat sinks. At this point, the teacher might ask how many animals were in the boat when it
sank. Children can then be asked to make predictions about how many animals it will take
to sink the boat on the next attempt. Each time, the predictions and actual results can be
charted. Teachers can guide children to refer to this chart when they make future predictions. The following questions and comments are examples of what teachers might say to
guide children's thinking:

- *Can you ever sink the boat with just one of these animals? Then we can infer that one of
  these animals is not enough to sink the boat.*
- *What number of animals came up the most often on the recording chart?*
- *How many animals do you predict that it will take to sink the boat this time?*
- *Betsy predicts three animals will sink the boat. Have three animals ever sunk the boat
  before?*
- *Takuo says that eight animals will probably sink the boat. How did you make that
  prediction, Tak?*
- *How many animals did the boat hold before it sank?*
- *What was the fewest number of animals that ever sank the boat?*
- *Why isn't the number of animals the same each time?*

Large-Group Activities

Large-group activities provide children with the opportunity to share ideas. They ensure that all children receive exposure to particular concepts, as some children who could not successfully engage in a particular activity by themselves may be able to participate at some level when they are part of a group. For example, children can often answer mathematical questions that are based on group graphs before they can graph data themselves. This is a good example of Vygotsky's (1978) ZPD. With teachers providing the necessary support, such as aligning the bars on a graph and describing their placement, children can begin to understand and analyze the data. Estimation and probability activities can also be successfully implemented at the large-group level.

Graphing Voting Data

Group time is a good venue for analyzing class voting experiences. As previously described, if children's names are listed on slips of paper that represent their vote on a topic, children can make a connection between what they voted for and the placement of their name in a certain column. Because older preschool and kindergarten children are increasingly interested in the ideas and decisions of their peers, group voting graphs are particularly appropriate for them.

There are several important criteria that teachers should consider when creating class graphs. First, the graphs should read from bottom to top. Young children are accustomed to judging magnitude based on height, so they quickly determine that the taller column on a graph has more items. Children may become confused when graphs start at the top, because all of the columns appear to be the same height. Also, when children are formally introduced to graphing in later grades, they learn to graph upward from the horizontal, *x*-axis. It makes sense to start this way in preschool and kindergarten.

Another important consideration when graphing is the placement of the bars, or tags, on the graph. Most young children do not understand the necessity of carefully aligning the tags in order to later analyze the data. They may leave large spaces between the tags and erroneously conclude that a tall column with a few tags that are spread apart represents more votes than a shorter column with more tags that are closer together. The best way to help children deal with this issue is to include grid lines on the graph. Children are less likely to create gaps on the graph when they can use one-to-one correspondence to fill in successive boxes. Teachers can also discuss the consequences of leaving gaps and demonstrate how the columns look when there are gaps versus how they look when there are no gaps.

A further consideration is the topic chosen for a graph. Teachers must be careful not to perpetuate stereotypes. In a math workbook sold at a national teachers' conference, children were asked to vote on whether they would weave rugs, make pottery, herd sheep, or make jewelry if they were Navajo. Topics such as this perpetuate stereotypes, in this case about the occupations of Diné, or Navajo, people. Another issue is the potential, hidden ramifications of some topics. For example, if teachers create a graph charting the types of pets that children have, some children may be upset because they do not have a pet and are not included on the graph. Good choices for graphing topics are those that relate to classroom circumstances, such as selecting a lunch menu, naming a class pet, or choosing a favorite animal after a zoo field trip.

A final consideration when creating graphs is the placement of a clearly marked label at the bottom of each column. Pictorial symbols, accompanied by written labels, help children identify what each column represents, while also supporting literacy development.

ACTIVITY 7.4

Author Unit Picks

Materials

The following materials are needed for this activity:

- Poster board or paper, 18 × 12 inches
- Teacher-made illustrations to depict the subject matter of selected books from the author unit
- Name tag for each child

Description

Preschool and kindergarten teachers sometimes highlight a particular children's author as part of an author unit. Various books by the author are assembled in the book area and read over the course of one or more weeks. A good way to conclude the author unit is to have children vote for their favorite book. To create labels for the voting columns on the graph, teachers should create illustrations that help children associate the columns with the particular books. Teachers can collect children's votes informally and then graph the results with the class. The names should be printed on the tags ahead of time to reduce wait time during the activity.

Favorite Book by Audrey Wood		
Mike		
Steve		
Chrissy		Carol
Aurora		Jay
Evan	David	Nancy
Claire	Rachel	Megan
Bela	Emily	Kevin
Isaac	Amanda	Heather
King Bidgood	Napping House	Silly Sally

Math Discussions

Children are likely to be interested in who voted for the same book as they did, and which book got the most votes. The conversation can begin with these topics. The teacher might ask, "How many people voted for the same book as Dwayne?" This encourages children to find Dwayne's name on the graph, perhaps with his help, and tally the number of votes in his column. After repeating this question with several names, the teacher might move on to comparisons among the books. Questions such as the following could lead that discussion:

- *Which book was the most popular?*
- *Which book had the fewest votes?*
- *Was there any book that nobody voted for?*
- *Did any books have the same number of votes?*
- *How many more votes did the winner get than the runner-up?*
- *If we voted again tomorrow, do you think the graph would look the same? Should we try it and see?*

Estimation

Estimation is another activity that works well within a large-group format. As children take turns creating estimates, they have the opportunity to compare their judgments with those of their peers. Teachers can further discussions about why children have made particular choices, and children can compare their estimates with the actual result. Skillful teachers can help their class judge which estimates are close to the actual amount. Counting up to and beyond a given value to determine whether an estimate is high or low may help children internalize the number line.

ACTIVITY 7.5

Comparing Apples and Oranges

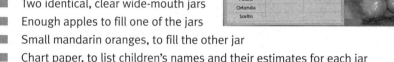

Materials

The following materials are needed for this activity:

■ Two identical, clear wide-mouth jars

■ Enough apples to fill one of the jars

■ Small mandarin oranges, to fill the other jar

■ Chart paper, to list children's names and their estimates for each jar

Description

To begin this activity, children examine the jar of apples and estimate how many it contains. Teachers can hold the jar so that it is clearly visible, or let children pass it around the circle. Even though many children will attempt to count the apples, they may still find it challenging to estimate how many are actually in the jar. As children take turns examining the jar, the teacher can write their estimates on the chart.

After each of the children has had a turn estimating the number of apples in the jar, the teacher can show the children the jar of mandarin oranges. An orange can be placed next to an apple so that the children can compare their sizes. Children can then take turns estimating how many oranges are in the jar. As teachers add each child's estimate of oranges to the chart, they can ask whether that child thought that there were more apples or oranges in the respective jars. This gives children the chance to make numerical comparisons. Those who have difficulty making these judgments can be assisted by their peers. Once all of the estimates have been completed, the jars can be opened and the contents counted.

Math Discussions

Many children will be surprised to discover that there are more oranges than apples in the jars. Children often expect there to be more of the larger item. For this reason, the math discussion should start with this topic. Children can offer ideas for why fewer apples will fit in the jar. Because the materials are available, the children can illustrate their ideas with the actual objects. Even after counting the apples and oranges, some children may not be convinced that there are more oranges. Teachers can line up the two fruits in a one-to-one correspondence relationship so that children can see the results. This configuration can also help children answer questions such as, "How many more oranges were in the jar than apples?"

Some children will want to return to this activity to see whether the results remain the same. The jars can be displayed in the classroom for several days so that children can experiment with the materials. If necessary, plastic apples can be substituted for real apples to discourage nibbling.

Probability Games

Carefully constructed probability activities can also be introduced during group time. Although children may initially make illogical decisions, such as basing their judgment on their favorite color, they enjoy participating in the gamelike probability activities. With experience, children may begin to understand some of the concepts related to likelihood. For example, if there are many blue blocks in a box and only one yellow block, even young children may notice that the blue blocks usually get picked.

ACTIVITY 7.6

Six Colorful Ducks

Materials

The following materials are needed for this activity:

■ Five plastic ducks, all the same color

■ One plastic duck that is a different color

■ Clear bag or other container, to hold the ducks

■ Tally sheet, with a picture of each color of duck, to record how many times a duck of each color is pulled from the bag

Description

Because probability concepts are difficult for young children, initial experiences should be as obvious as possible. For this reason, the ducks in this activity are placed in a container that allows them to remain visible to the children. Even young children can see that there are more ducks of one color than of the other. The ducks can be counted as they are placed in the bag. Children can then take turns coming to the front of the group, closing their eyes, and pulling a duck from the bag. Before they select a duck, children should predict the color of the duck they will pull from the bag. The teacher can place a mark beneath the appropriate color of duck on the tally sheet to record the results. After each turn, the duck should be returned to the bag.

Ducks were selected for this activity because they are readily available and children find them interesting; however, colored balls, blocks, or other objects can be substituted. This activity can be repeated and varied over time. Perhaps, a second duck of the alternate color can be added to the bag and children can compare the outcomes.

Math Discussions

Throughout the Six Colorful Ducks activity, children should be encouraged to talk about their predictions and compare the outcomes. Periodically, the teacher and children can count the tally marks and compare the totals for the two colors of ducks. The teacher may ask children if they want to consider the results of the previous attempts before they make their predictions; however, they should be prepared for illogical answers, such as, "I picked blue because that's the color I want to get." Vocabulary related to probability should be incorporated into the conversation. Here are some examples:

- *Do you think it is very likely that Larry will pull a blue duck from the bag? There's only one in there.*
- *The last three people picked a yellow duck. Is it possible to pull a blue duck out of the bag?*
- *What could we do to make it more likely to pull a blue duck out of the bag?*

Data Analysis and Probability Throughout the Curriculum

In order for children to become familiar with the terminology associated with data analysis and probability, they need to hear the vocabulary often and within an understandable context. Data sources are located throughout the classroom in the toys, art materials, activities, and foods that children experience. Teachers should use these areas of interest to introduce meaningful comments, questions, and problems related to data analysis, estimation, prediction, and probability. The examples that follow illustrate how these concepts can be integrated throughout the classroom.

Lunch or Snack

Some foods are mixtures of two or more items. Examples include peas and carrots, corn and lima beans, rice and beans, and mixed cereals. These foods provide a natural data source for children to analyze, and encourage them to make predictions. As an example, if peas and carrots are served, there will undoubtedly be many more peas than carrots, although the carrots are larger. Teachers can ask children whether they have more peas or carrots on their plates. Before the next helping is served, teachers can review the data (e.g., everyone had more peas) and ask children to predict whether they will have more peas or carrots this time. This type of conversation can occur frequently during lunch and snack time.

Sensory Table

Teachers sometimes add small aquatic animals to the water table for children to play with, perhaps in conjunction with a popular story. In the book *King Bidgood's in the Bathtub*, by Audrey Wood (1985), the king goes fishing in the bathtub. As an extension of the story, children can use small nets to fish for toy turtles, frogs, fish, and plastic worms in the water table. A board stretched across the water table provides a place for children to sort the contents of their fishing nets. Teachers can make comments or ask questions that encourage children to compare their groups, and follow-up questions can relate to probability. For example, if a child caught many fish, but only one turtle, the teacher might ask the child to predict whether there will be more fish or turtles in his next attempt.

Art Area

A well-stocked art area is full of data sources. Some teachers display collage materials in divided trays so that children can easily see the items. In a sense, this is a collection of data that have already been sorted. As an example, a spring collage tray might include pastel cotton balls, pieces of ribbon, and feathers. During a conversation about the materials, teachers might ask whether the children think that there are as many cotton balls as ribbons. Children may erroneously assume that there are more cotton balls because they take up a lot of space in their compartment, especially when compared with ribbons, which lie flat. Math problems such as this connect data analysis to measurement and number.

Manipulative Area

The manipulative area is filled with opportunities for data comparisons. Because children often like to hook manipulative pieces together, teachers might suggest that two children create rows of the same length, using different materials. They can then predict whether the rows will have the same number of pieces, or whether one row will have more. When children string beads, the teacher might hold in his hand several beads of one color and one bead of another color. Children can predict which color of bead they will pick if they close their eyes and select one from the teacher's hand. When children are playing with pegboards, the teacher might ask whether there is the same number of each color of peg for the board. The opportunities to compare sets and make predictions are endless.

Science Area

In the science area, collections of items such as shells, rocks, pinecones, or leaves are often displayed. Teachers can encourage children to sort the materials in different ways and compare the groups. When children explore physical-knowledge materials, such as ramps, scales, pulleys, levers, or pendulums, teachers can suggest that the children predict what will happen before they try a new experiment.

UNIVERSAL DESIGN FOR SUPPORTING DATA ANALYSIS AND PROBABILITY

All children can participate at some level in data analysis and probability experiences. In the area of data analysis, sorting and classifying activities can be quickly modified to meet individual needs. For example, children with cognitive delays or short attention spans may initially need collections with fewer items and one strong attribute, such as color. Teachers can gradually add more items and a second attribute, such as size. For children who have language delays or are learning English as a second language, gestures that accompany mathematical language may be helpful. For example, when size is the sorting criteria, teachers can gesture for big and little while also saying the words. Children who are blind or have severe visual disabilities need data that can be organized by the use of tactile or auditory information. A teddy bear collection that includes furry, felt, plastic, and wooden bears of various sizes is an example. Another possibility would be a collection of jingle bells and cow bells of different sizes. The two types of bells have different tone colors, and the size alters their pitch. This gives children two types of sound stimuli to use as sorting criteria.

Children use strategies commensurate with their level of quantification to compare groups of objects. For example, children at the global level make judgments based on how large each group looks. Teachers can move them forward by modeling one-to-one correspondence, in which items from each group are paired. If one group has objects left over, children can begin to understand that it has more. Naturally, if children are already at the one-to-one correspondence level, teachers will want to model counting.

Making predictions is challenging for most young children, but all should be given the opportunity to participate. When implementing probability activities such as Six Colorful Ducks, teachers can use gestures and pantomime to show children with cognitive and language delays the sequence of the activity. The children can point to the color they choose.

As always, the teacher should make accommodations to activities before they are introduced to the class. For example, if a child in the class is color blind, then colors that are not problematic should be selected for the activities. Black and white objects can be used for all children when the class includes children who need high contrast in order to maximize their residual vision. If a blind child is included in the class, then a tactile element should be added to all materials before they are introduced. For the duck example, feathers could be added to one color of duck so that it can be distinguished by feel from the other color of duck.

Children who have very short attention spans may participate in individualized programs that limit their time at large-group experiences. They can be among the first (but not always the first) children called for a turn during math activities. They may also be able to return to the group for some of the follow-up discussion.

FOCUSING ON MATHEMATICAL PROCESSES

- *Problem Solving*—This standard is used when children or teachers pose questions that can be answered by data. To answer the question, the children must collect and analyze the data and interpret them to one another.

- *Reasoning and Proof*—Children use this standard when they attempt to explain to one another their interpretations of graphs. They may use visual comparisons, one-to-one correspondence, or counting to justify their responses.

- *Communication*—As children determine attributes for sorting and classifying data, communication is strongly involved. Children also use language to analyze data on graphs and communicate the results to others such as their parents, and to communicate their reasoning when they make predictions.

- *Connections*—Data Analysis is related to all of the other mathematics content standards. Focusing on attributes through sorting and classifying is also part algebra, as children must distinguish attributes before they can use them to create patterns. Comparing and quantifying data connects to number sense and measurement. Because shape is a primary attribute of objects, data analysis also connects to geometry.

- *Representation*—Graphing is the main focus of representation in the Data Analysis and Probability standard. In preschool and kindergarten, most of the graphs are group representations, such as voting data. Completing graphs as part of a group allows teachers to provide the necessary scaffolding for young children.

ASSESSING DATA ANALYSIS AND PROBABILITY

Anecdotal notes are the best way to assess children's understanding of data analysis and probability during preschool and kindergarten. Children's thinking is revealed through the conversations that teachers lead during individual, small-group, and large-group experiences, and these conversations can be recorded. A small notebook that fits into a pocket is a convenient tool to use to quickly capture children's responses. Using children's initials and the teacher's own personal shorthand, the teacher can quickly note what children say and do. If children ask what the teacher is writing, a general, yet truthful, response can be provided. For example, the teacher might say, "I'm writing down what we did today so I don't forget," or "Everyone has such interesting ideas. I want to remember what everyone says about this." The notes can be transcribed into the children's classroom files or portfolios, or into a computer file at the end of the day while the details are still fresh in the teacher's mind.

Sometimes, children create a mathematical model that is so complex that a photograph is needed to capture it. Figure 7.7 is such an example. A gifted 5-year-old organized a pegboard to reflect complex symmetrical relationships. Every figure on the pegboard was carefully placed in accord with this pattern. For example, farmers were placed at the four corners of the pegboard, and each farmer had a white sheep facing him. Along opposite edges of the pegboard were identical rows of black animals that also had an internal symmetry (horse, sheep, horse, horse, horse, horse, sheep, horse). Along the other two edges of the square were identical rows of pigs that were, again, internally symmetrical (white, black, black, black, black, white). The child's organized vision encompassed the entire pegboard design. It is difficult to capture an image such as this in words, but a photograph quickly preserves it for later reflection and discussion with the child.

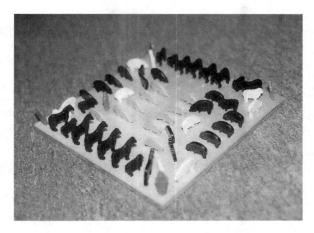

Figure 7.7. Photograph of symmetrical pegboard design.

SUMMARY

The Data Analysis and Probability standard focuses on posing questions that can be answered by data, drawing conclusions from the data, and making predictions on the basis of those data. Before data can be analyzed, they must be collected, organized, and displayed. For young children, materials and experiences from everyday life provide data to analyze. They may sort toys or other objects into groups and make comparisons on the basis of those categories. Children also like to vote on issues that concern the classroom, and they evaluate voting data that are displayed on group graphs.

A second component of this standard is beginning concepts about probability. Although most preschool and kindergarten children have not yet developed the reasoning skills to make probability-related choices, they can begin to consider concepts such as "likely or unlikely" and "always or never." Estimation is a concept that is related to probability. With experience, young children can begin to make estimates with some degree of accuracy for amounts of less than 10 units.

The Data Analysis and Probability standard incorporates concepts from the other four mathematics content standards. When children sort objects into categories, they must form general rules to distinguish the groups. This focus on attributes and overarching rules is connected to the Algebra standard in the early years. Children compare groups of objects depending on their level of quantification: global, one-to-one correspondence, or counting. This is a major component of the Number and Operations standard. Shape and size are two of the attributes of objects that children use to form groups. Focus on these two attributes connects Data Analysis to both the Geometry and Measurement standards, respectively. In addition, the analysis of bar graphs is closely connected to measurement. Children rely on the height of the columns to make judgments; however, as with measurement, they must be guided to also focus on the size and placement of the units.

Teachers can help children construct concepts related to Data Analysis and Probability by making relevant comments or posing related questions throughout the day, in all areas of the curriculum. In addition, teachers should plan regular small- and large-group activities that focus on these concepts.

ON YOUR OWN

■ Compile a list of children's books that contrast fantasy and reality, such as teddy bears versus real bears.

■ Think of three topics that would each make a good voting graph for young children.

■ List a source of data that children could organize in each area of the classroom.

REFERENCES

Allen, P. (1996). *Who sank the boat?* New York: Putnam Juvenile.

Baratta-Lornton, M. (1976). *Mathematics their way.* Menlo Park, CA: Addison-Wesley.

Baroody, A.J., Lai, M., & Mix, K. (2006). The development of young children's early number and operation sense and its implications for early childhood education. In B. Spodek & O.N. Saracho (Eds.), *Handbook of research on the education of young children* (pp. 187–221). Mahwah, NJ: Erlbaum..

Clements, D.H., & Sarama, J. (2007). Early mathematics learning. In F.K. Lester, Jr. (Ed.), *Second handbook of research on mathematics teaching and learning* (pp. 461–555). Reston, VA: National Council of Teachers of Mathematics.

Jones, G.A., Langrall, C.W., & Mooney, E.S. (2007). Research in probability: Responding to classroom realities. In F.K. Lester, Jr. (Ed.), *Second handbook of research on mathematics teaching and learning* (pp. 909–955). Reston, VA: National Council of Teachers of Mathematics.

Kamii, C. (1982). *Number in preschool and kindergarten: Educational implications of Piaget's theory.* Washington, DC: National Association for the Education of Young Children.

Kamii, C. (2000). *Young children reinvent arithmetic: Implications of Piaget's theory* (2nd ed.). New York: Teachers College Press.

Metz, K.E. (1998). Emergent understanding and attribution of randomness: Comparative analysis of reasoning of primary grade children and undergraduates. *Cognition and Instruction, 16,* 285–365.

Moomaw, S., & Hieronymus, B. (1995). *More than counting.* St. Paul, MN: Redleaf Press.

National Council of Teachers of Mathematics. (2000). *Curriculum and evaluation standards for school mathematics.* Reston, VA: Author.

National Council of Teachers of Mathematics. (2006). *Curriculum focal points.* Reston, VA: Author.

Piaget, J. (1965). *The child's conception of number.* New York: Norton.

Piaget, J., & Inhelder, B. (1976). *The origin of the idea of chance in young children.* New York: Norton.

Shaughnessy, J.M. (1992). Research in probability and statistics: Reflections and directions. In D.A. Grouws (Ed.), *Handbook of research on mathematics teaching and learning.* Reston, VA: National Council of Teachers of Mathematics.

Vygotsy, L.S. (1978). *Mind in society: The development of higher psychological processes.* Cambridge, MA: Harvard University Press.

Wadsworth, B. (1989). *Piaget's theory of cognitive and affective development* (4th ed.). White Plains, NY: Longman.

Wood, A. (1985). *King Bidgood's in the bathtub.* San Diego: Harcourt Brace Jovanovich.

8

Integrating Curricula to Meet Mathematics Goals

Putting It All Together

One of the most widely accepted ideas within the mathematics education community is the idea that students should understand mathematics. . . . But achieving this goal has been like searching for the Holy Grail. There is a persistent belief in the merits of the goal, but designing school learning environments that successfully promote understanding has been difficult.

—James Hiebert and Thomas P. Carpenter (1992)

This textbook has focused on the importance of the five mathematics content standards developed by the NCTM (2000) to the education of young children. A number of suggestions for promoting an interactive mathematics curriculum at the individual, small-group, and large-group levels have been presented, along with many specific examples. The challenge for preschool and kindergarten teachers is to implement such a comprehensive mathematics curriculum in a way that meets the learning needs of all students and promotes the construction of foundational mathematics concepts. Planning, scheduling, and implementing the mathematics curriculum is the goal of this chapter.

Preschool and kindergarten teachers work within a variety of programs, including public, private, full-day, half-day, and year-round schools. Some kindergarten teachers have a general scope and sequence for mathematics topics, whereas others have a specific, adopted curriculum. In addition, most kindergarten teachers, and many preschool teachers, have state-adopted benchmarks and indicators for mathematics content standards. How can a current or prospective early childhood teacher prepare to teach mathematics under such diverse circumstances?

Fortunately, most states that have adopted mathematics standards have relied extensively on the NCTM standards. Teachers who understand this content relative to preschool and kindergarten age groups are well prepared to implement a comprehensive mathematics program regardless of their geographic area. Second, it is important to remember that all teachers can integrate math talk into their classrooms throughout the day. Teachers who feel squeezed for time due to a busy schedule can take advantage of transitions and other daily activities to interject math topics. Quietly chanting rhythmic patterns while waiting for the bathroom, looking for geometric shapes while en route to the playground, sorting and seriating materials during clean up, and generating math problems during snack and lunch are just a few of the activities that take advantage of daily opportunities to integrate math. Finally, all teachers, including those with an adopted curriculum, must make adaptations to meet the learning needs of individual children. Many teachers express frustration when they are required to implement a particular activity in a given way, because they know that the children do not understand the concepts. In these situations, teachers need to go beyond the regular curriculum and find other ways to introduce and follow up on the same concepts. Mathematics is too important to let learning languish due to a sterile curriculum. By planning additional activities for small and large groups, along with center-based activities, teachers can cover required content while supporting the conceptual development of children through an augmented curriculum.

The guidelines and scheduling suggestions that follow are provided with all teachers in mind. Teachers who are free to develop their own curriculum will find a framework for long-range planning and implementation. Teachers who have a required curriculum can use the ideas to augment their adopted curriculum. Because many preschool and kindergarten programs have half-day schedules, the curriculum frameworks are directed toward part-time schedules. Teachers of full-day programs have the option of introducing more activities or expanding the schedules across a day-long time frame.

GUIDELINES

A robust mathematics curriculum for preschool and kindergarten encompasses many inter-related concepts. What content to include, when to introduce it, and how to implement the curriculum are just a few of the questions that teachers must address. The 10 guidelines that follow summarize the important components of a comprehensive mathematics curriculum for young children.

Guidelines for Designing and Implementing Mathematics Curriculum

1. *Mathematics curricula for young children should be developmentally appropriate.* Teaching practices should be appropriate for each child's age and developmental level, and they should be based on knowledge about how children develop mathematical concepts (Copple & Bredekamp, 2009).

2. *All five mathematics content standards should be included.* Although Number and Operations, Geometry, and Measurement are focal areas for early childhood mathematics (NCTM, 2006), Algebra, and Data Analysis and Probability are also important and support learning in the other mathematics content areas.

3. *Teachers should develop long-range goals for each area of the mathematics curriculum.* These goals should be based on the developmental levels of the class, and they should include planning for individual children as well as the group. Longitudinal planning ensures that the important concepts in each area are covered.

4. *Accommodations should be made to the curriculum so that all children can participate to their maximum potential.* These accommodations should be made before the curriculum is introduced.

5. *Teachers should develop a regular schedule of implementation for all areas of mathematics at the individual, small-group, and large-group levels.* Implementation can occur during choice time, special activities, regular small-group work, or larger group times.

6. *Special emphasis should be placed on the development of number sense.* Some researchers believe that the concepts embedded in number sense are as important to early mathematics learning as concepts of phonemic awareness are to early reading (Gersten & Chard, 1999).

7. *Teachers should make every effort to include the language of math in daily conversations with children.* Research shows that the amount of math-related talk that teachers provide is significantly related to children's mathematical learning (Klibanoff et al., 2006).

8. *Mathematics should be integrated throughout the curriculum.* An integrated curriculum allows children to make important learning connections and increases their exposure to mathematics.

9. *Teachers should be as familiar with the mathematics process standards as they are with the content standards.* Use of mathematical processes is critical to children's understanding of mathematics concepts. The process standards should be integrated routinely into children's mathematical experiences.

10. *Assessment should be an ongoing part of instruction* (NCTM, 2000). Teachers should use assessment to scaffold children's learning. Children's responses in mathematical situations should guide teachers on what to say, how to model at the appropriate level, and what additional activities to plan.

In the sections that follow, these 10 guidelines are incorporated into specific planning for each of the mathematics content standards.

PLANNING FOR NUMBER AND OPERATIONS

There are four critical areas related to number and arithmetic operations for young children: quantification concepts, counting skills, representation of number, and emergent addition and subtraction operations (see Chapters 2 and 3). These areas should be the focus of long-range planning in preschool and kindergarten. Short-range planning should support children's current levels of development in these areas and challenge them just enough to move forward in their thinking. Assessment is an ongoing part of planning and implementation. Through assessment, teachers know not only what activities to plan, but also how to modify them during implementation in order to maximize learning.

Longitudinal Planning

Longitudinal planning should encompass the developmental sequences of learning in quantification, counting, representation, and addition or subtraction. Initially, this planning is based on anticipated learning trajectories, given the ages and previous experiences of the children. During the course of the year, long-range goals may change somewhat depending on the progress of the children. Nevertheless, long-range planning is important because it provides a framework for teachers to ensure that important concepts are adequately addressed.

Developmental Sequences for Preschool

Many mathematical concepts follow a developmental sequence. Understanding of this progression allows teachers to support children at their current level of thinking and move them toward the next anticipated level in their learning trajectory. Consequently, planning must reflect the various developmental levels present in any classroom. **Self-leveling** activities, such as manipulative materials and board games, allow children at various stages of quantification and counting to participate together and move forward in their thinking.

Quantification   Preschool teachers can expect children to be at all three levels of quantification—global, one-to-one correspondence, and counting—at the beginning of the school year. For this reason, math manipulative games and board games that include a 1–3 or a 1–6 dot die or spinner should be included from the beginning of the school year. Teachers of younger preschool children may begin with manipulative and grid games, as these materials are most likely to accommodate the thinking levels of 3-year-olds. Teachers of older preschool children, however, should have short- and long-path games available for children who are already at more advanced levels of quantification and counting. Math games are invaluable to teachers for making initial assessments of children's thinking and for guiding subsequent planning. Within just a few minutes, teachers can accurately gauge children's quantification levels and their understanding of counting principles.

Throughout the school year, teachers should expect children to move from simple manipulative and grid games to short and then longer path games, which are conceptually more abstract. They should also expect children who begin at the global stage of quantification to progress to one-to-one correspondence and counting, at least for small quantities. Children who begin school already at the counting stage should progress to quantifying larger quantities and eventually to combining two dice (adding) when playing math games.

Counting   Counting should be a regular focus of the preschool curriculum throughout the year. Group-time counting activities, such as songs and fingerplays, allow teachers to model important counting principles, including stable order counting, a one-to-one correspondence relationship between counting words and objects being counted, and cardinality. As children develop counting concepts, considerable variation in skills may be apparent. For example, some children may be proficient at stable order counting, but may not understand cardinality; therefore, they cannot use their counting skills to quantify objects. Other children may understand cardinality, but leave out a counting word or mix up the order of the count string. Most preschool children skip over or re-count some objects, at least some of the time. As a general guideline, teachers should plan group-time counting activities for sets of five at the beginning of the year. (See Figure 8.1 for examples.) As the year progresses, teachers can move to sets of 10 and, perhaps, sets of up to 20 by the end of the school year, especially for classes of older preschool children. At the individual level, widespread differences in counting may be apparent. Teachers should support each individual child at his or her current level of understanding, always with the goal of moving each child a step farther in development.

Addition and Subtraction   Problem-solving situations, which often involve addition and subtraction, should be a goal of planning from day one. Teachers should anticipate opportunities to introduce problem solving throughout the classroom, as well as during group time. At first, the mathematical situations should involve very small numbers, such as adding or subtracting one item from a group of five or less. For example, in dramatic play the teacher might say, "I have two eggs. Can I have one more? Now how many do I have?" During group time, the teacher might introduce and model a traditional number song, such as "Ten in the Bed," but reduce the quantity to five. This introduces children to the idea of subtracting by one. Later, more complex problems can be introduced. No one can preplan this sequence for teachers, as it will depend on each individual class. With experience, teachers learn which activities work especially well at various times of the year. The best guideline for beginning teachers is to select an activity that they think will interest their class. If it is too difficult, try it again with smaller numbers. If it is too easy, move on to higher numbers.

> **"Five Little Leaves" (traditional fingerplay)**
>
> Five little leaves so bright and gay,
> Were dancing about on a tree one day,
> The wind came blowing through the town,
> And one little leaf came fluttering down.
>
> *(Repeat until no leaves are left.)*
>
> **"Five Little Pumpkins" (by Sally Moomaw)**
>
> Five little pumpkins growing on a vine,
> Mr. Raccoon said, "This one's mine!"
> He pulled one pumpkin from the patch,
> Said, "Mmm, mmm, mmm, what a fine catch."
>
> *(Substitute various animals into the verse until there are no pumpkins left.)*

Figure 8.1. Sample fingerplays for counting.

Teachers can also introduce addition and subtraction through math games (see Chapter 3). At first, children may count separately the amounts on two dice. Next, they may count the amounts separately, but then re-count them all together—a *count-all* strategy. Some children quickly move to a *short-cut sum* strategy, in which they immediately count both sets of dots all together. When children have had many opportunities to add two dice, they often begin to remember combinations, beginning with the doubles and plus one. Some children also begin to *count on* from the perceived cardinal value of the first die, although this usually does not occur during preschool. These progressions in addition strategies unfold rather naturally in classes in which children have many opportunities to play math games. Long-range planning, then, requires assembling games of sufficient complexity and interest to keep children motivated.

Representation The primary method that preschool children use to represent mathematical problems is to model them with real objects—play materials, fingers, counters, and so forth. Children also use drawings to show mathematical situations, such as drawing a family of four or illustrating two friends. When keeping score, children sometimes use self-invented symbols to represent quantities, such as hash marks or circles for each point earned.

Teachers should model writing numerals to represent quantities in appropriate situations. For example, if a class has just calculated that five children have returned permission slips, the teacher could use a numeral when writing that message. It is important to remember, however, that research shows that even when children can read and write numerals and know the quantities they represent, they seldom use numerals when representing mathematical situations (Kato, Kamii, Ozaki, & Nagahiro, 2002).

Developmental Sequences for Kindergarten

Just as preschool children vary widely in their understanding of mathematical concepts, kindergarten children enter school with very different experiences and knowledge of numbers. Whereas kindergarten teachers can expect their classes to begin at a higher math level than preschool, they may find that they need to backtrack for some children during the beginning weeks of school. The important goal for teachers should remain to support children at their current level of understanding, but to hold high expectations for their ability to grow during the school year.

Quantification and Counting Kindergarten teachers can expect many of the children in their classes to already be at the counting stage of quantification when school begins;

however, children who have not been to preschool or who have had limited experiences with math may employ one-to-one correspondence or even global strategies when quantifying. Even though they may be able to rote count, some children may not yet understand the cardinality principle when counting. For this reason, the beginning kindergarten curriculum should focus heavily on opportunities to count so that teachers and peers can model cardinality.

Once again, math manipulative and board games strongly support learning in this area. Kindergarten teachers may choose to incorporate them during small math groups as a regular part of the daily schedule. These groups should be flexible. Sometimes, children at approximately the same developmental level may be grouped together so that the materials can be gauged to meet their needs (e.g., quantity of dots on dice, number of counters). At other times, mixed groupings should be incorporated. Children at higher levels of development provide excellent models for their peers and can often explain their strategies and why they work. This is beneficial for all children. Some children have the support of a peer-teacher, whereas more advanced students must reflect on their thinking in order to explain it to others.

As with preschool, counting should be incorporated regularly into classroom routines and group times. Counting is such an important foundational math concept that repeated reinforcement is desirable. Once children are proficient at counting forward, backward counting can be introduced. Backward counting helps some children envision subtracting along a number line. As the year progresses, some kindergarten teachers may move into counting by twos and fives. Counting by twos can emerge naturally when children line up in two lines. At first, the teacher may count every student, but emphasize the even numbers. Later, the teacher can model counting by twos. Counting by fives can by introduced when objects have been grouped into sets of five. The objects can first be counted individually, with an emphasis on every fifth counting word. Then they can be recounted as groups of fives—5, 10, 15, and so forth. Many kindergarten classrooms participate in their school's Day 100 celebrations. Preparation for these activities gives children experience in counting through the decades.

Addition and Subtraction   Math games such as those suggested in Chapter 3 help kindergarten children develop beginning addition concepts and also some fluency with addition of small numbers. The same progression in combining sets as that described for preschool should unfold in kindergarten, although some children may begin at a higher level. For example, rather than counting all, some kindergarten students may already employ a short-cut sum strategy. Many kindergarten children may begin to count on from the first addend to reach the total during the course of the year. If quantities are represented by numerals, children may need to use their fingers or some type of audible marker, such as taps, to represent the group that they are adding. Many kindergarten children may be able to add two quantities when concrete symbols, such as dots, are used to represent the groups, but be unable to do so when numerals are used. They may add the numeral as "one" regardless of its cardinal value, or simply ignore it. For this reason, many children may need to spend a long time adding concrete sets, such as dice with dots, before they begin adding numeric representations.

Long-range planning in kindergarten should include many opportunities for children to compose and decompose numbers. At first, these activities should involve small groups of about three to five objects. Gradually, larger amounts can be introduced. Composing and decomposing numbers helps children understand the relationship between addition and subtraction, as well as remember basic addition and subtraction combinations.

Short-Term Planning

Short-term planning focuses directly on the current understanding and interests of the children. Skillful teachers often coordinate math games and materials with topics or books of interest. This helps children to connect mathematics to other areas of the curriculum and often increases their interest in math materials.

Planning for Preschool

In preschool, multiple quantification games should be available daily for children to use during choice time, which accounts for a large part of the schedule in most preschool programs. Manipulative materials and board games can be displayed on manipulative shelves or a special game table. In most classes, materials to accommodate a range of developmental levels will be necessary.

Number and operations should be incorporated at least twice a week into group time. This could be through number songs, interactive charts, books, dramatizations, and so forth. Also, group-time activities from other math content standards often contain number concepts. For example, children might count the sides of polygons or graph and count voting data.

When setting up the classroom, teachers should plan to integrate numbers into at least one other area of the classroom. For example, one-to-one correspondence could be integrated into the water table, or quantification could be a planned part of the dramatic play area. See Chapters 2 and 3 for many ideas for an integrated curriculum.

Finally, number sense can be the focus of a weekly special activity. The special activity does not always have to be art! Math games, math-related art projects, and class books all make interesting special activities.

Planning for Kindergarten

Kindergarten programs should also have many math materials and games available for daily explorations. These materials can encourage quantification, counting, composing and decomposing number, seriation of numbers, addition, and subtraction. As suggested earlier, kindergarten teachers should also plan a daily math experience. This might be introduced to the whole group and then completed in small groups. Opportunities to play quantification games should be a priority. Activities that focus on composing and decomposing numbers, representing number in various ways, and creating math-related class books can also be planned.

Numbers should be a daily part of kindergarten group times. The same variety of activities suggested for preschool are also appropriate for kindergarten, although they can be approached at a higher level. For example, kindergarten children can sing songs that involve addition and subtraction of small numbers or help the teacher model math problems that occur in story books. Counting in a variety of ways, such as forward, backward, or by twos, can also be incorporated into group-time activities.

Like preschool teachers kindergarten teachers should plan ahead for ways to incorporate math into other areas of the curriculum. With experience, dedicated teachers can become proficient at interjecting math into various subject areas and noninstructional times, such as transitions. In the beginning, though, teachers may find it helpful to include these opportunities in their lesson plans so that they do not forget to incorporate them.

PLANNING FOR ALGEBRA

Three areas are directly related to algebra in preschool and kindergarten: sorting and classifying, patterning, and modeling mathematical situations. These three areas should be a focus of long-range planning so that they are adequately represented in short-term planning.

Longitudinal Planning

Algebra concepts support number sense by encouraging children to create mathematical relationships that can then be compared. For example, after sorting a group of objects into categories, children may then compare them to decide which group has the most, the fewest, and so forth. When creating patterns, children must focus on how many iterations of each attribute occur within a repeating sequence, such as "big, big, little, little." Algebra activities therefore help children connect number concepts to new mathematical situations. For this reason, algebra activities should be well represented in longitudinal planning.

Developmental Sequences for Preschool

As with number sense, there are developmental sequences to children's understanding of algebra concepts. These are mainly related to the complexity of the task, such as the type and number of attributes that a child must consider when sorting objects into groups or incorporating them into patterns. The ability to model mathematical problems may be affected by the size of the numbers included in the problem, the number of steps involved, and the mathematical operations that may be necessary.

Sorting and Classifying   Sorting and classifying requires children to focus on a particular attribute of a group of objects in order to place them in various categories. In other words, an overriding rule determined by the child, or in some cases by the teacher, determines which objects can fit into a category and which cannot. Because children must be able to identify particular attributes before they can use them to create patterns, sorting and classifying experiences should precede patterning activities that involve objects.

Collections of objects for children to sort by various attributes should be an ongoing part of the preschool mathematics curriculum. For introductory experiences, or in activities planned for young children, these attributes should be easily identifiable. For example, young children often group objects by color or type. As the year progresses, collections of objects with more attributes or finer detail can be introduced. In general, children first focus on only one attribute in a collection, although more than one attribute should be present. With experience and teacher input, children learn to sort collections by various attributes. Eventually, some preschool children may be able to sort items on the basis of two simultaneous attributes, such as size and color.

A well-equipped preschool mathematics curriculum should always have an interesting collection available for children to sort. Often, these collections can be related to a topic of interest. For example, if a zoo field trip has been planned for April, then the longitudinal plan may suggest a collection of zoo animals to sort and classify for that month as well. This does not mean that teachers can or should plan their entire curriculum at the beginning of the year. The complexity of the collection that is introduced will depend on the level of understanding of the children in the class at the time that the collection is implemented. Sorting and classifying can also be introduced occasionally during group-time experiences.

Patterning   Patterning is an important mathematical concept that provides a framework for understanding the number system and solving many types of problems. The first patterns that preschool children are likely to understand and extend are those that involve rhythm, music, or movement. Patterning activities that include these elements can be introduced early in the school year. For example, children can clap the rhythm of their names, which creates a pattern when repeated (e.g., Mi-chelle, Mi-chelle, Mi-chelle). They can also recognize and repeat patterns in songs, such as "B-I-N-G-O" or "E I E I O" in "Old MacDonald."

After children have had experiences sorting and classifying objects according to visual and tactile attributes, patterns based on these elements can be introduced. The easiest are alternating patterns that involve one attribute, such as red-blue, red-blue. As the year progresses, preschool teachers can add a third attribute to patterns, such as dog-cat-rabbit, dog-cat-rabbit. The difficulty of the patterns should depend on the age of the class, the number of experiences provided, and the speed at which children construct pattern relationships. In general, goals for preschool include recognizing, extending, and creating simple patterns, such as those previously suggested.

Modeling Mathematical Situations   Mathematical situations can be created by the teacher to reinforce understanding and representation of specific concepts. Because these math problems are often closely tied to number-sense concepts, they should be based on the child's understanding of quantification and counting. Teachers may introduce math problems based on a children's book, or they may watch children play and then interject a math-related situation. At the beginning of the year, teachers might encourage children to model the sequence of characters in a predictable book, such as *I Went Walking*, by Sue Williams (1992). As the year progresses, more difficult situations can be introduced. For example, children might recreate the scenarios in Pat Hutchin's book *1 Hunter* (1986).

Developmental Sequences for Kindergarten

Developmental sequences, as reflected in kindergarten planning, are similar to those suggested for preschool. As always, the learning trajectories will build upon the experiences that children have had prior to kindergarten. In general, kindergarten classes can be expected to progress more quickly and reach more advanced levels within the various areas of algebra.

Sorting and Classifying   Kindergarten children are usually very interested in sorting and classifying activities. Because they are older, they may be more flexible in grouping items by various attributes than preschool children are. For this reason, kindergarten teachers can introduce larger and more complex collections. As the year progresses, kindergarten children may be able to distinguish items that belong in more than one category. For example, if the sorting categories are blue and red, items that have both colors on them may be placed in a group between the two colors.

Kindergarten children may extend their sorting of items to include classification hierarchies. For example, dinosaurs might be grouped into meat eaters and plant eaters, and then be further separated by type of dinosaur within those groups. Also, kindergarten children can extend their sorting activities into other mathematics domains. For example, they might graph a collection once it has been sorted and compare the quantities in the various groups. Long-range planning for kindergarten should include the extension of sorting experiences into related areas.

Patterning   Some children will enter kindergarten with a beginning, conceptual understanding of patterning, whereas other children may quickly construct fundamental patterning concepts. For this reason, kindergarten teachers should implement simple patterning activities near the beginning of the year, assess children's understanding of patterning concepts according to these introductory activities, and move forward on the basis of that information. Throughout the year, patterning should be introduced in many different contexts: through movement, rhythm, music, geometry, science, and art. Longitudinal planning can help teachers organize the inclusion of patterning in various areas of the curriculum, as well as the introduction of more difficult patterns (see Chapter 4).

Modeling Mathematical Situations   Kindergarten children need many opportunities to model mathematical problems. These can come from daily classroom situations, such as planning food needed for a class party, or can be introduced through children's literature.

Teachers may find it easier to focus on opportunities for children to model mathematical situations when they are completing short-range plans. However, some opportunities may fit in particularly well with school celebrations or planned topics of interest and, therefore, may be included on long-range plans as well. For example, teachers might want children to create mathematical models for the book *One Hundred Hungry Ants,* by Elinor J. Pinczes (1999), when the school is celebrating the first 100 days of the year, or when the class is planning a year-end picnic. Particular activities such as this can be slotted into the longitudinal plan for the year.

Short-Term Planning

Through short-term planning, teachers can design collections, patterning activities, and opportunities for representation according to children's developmental levels, current interests of the children, or class topics of interest. For example, if a class is interested in autumnal changes, teachers might assemble collections of pine cones, seed pods, leaves, or nuts for sorting and classifying, as well as patterning activities.

Planning for Preschool

Collections of objects to sort and classify should be available regularly in preschool classrooms. They can be displayed along with a sorting tray in the manipulative area, or placed on a small table or bench. Collections can also be integrated into other areas of the classroom. For example, a collection of plastic fish could be sorted in the water table, and a collection of sea shells could be placed in the science area.

Patterning activities may be initially incorporated as part of group-time experiences, particularly if they involve rhythm, music, or movement. As the year progresses, patterning experiences that incorporate objects can be planned for the manipulative, dramatic play, and art areas. Once children have constructed fundamental patterning concepts, patterning should become a regular part of the curriculum so that children can build upon these initial concepts. Opportunities to model mathematical situations should be introduced into children's play throughout the classroom on a regular basis.

Planning for Kindergarten

Kindergarten children should have regular opportunities to sort and classify materials and create patterns. Collections of objects can be available daily during choice time and incorporated weekly into small math groups for sorting or patterning activities. In addition, kindergarten teachers should regularly incorporate patterning into group-time activities. Teachers can introduce mathematical problems for children to model, develop problems to accompany daily activities, or connect mathematical modeling to children's literature.

PLANNING FOR GEOMETRY

Three areas of geometry are important to include in planning the mathematics curriculum: two- and three-dimensional geometric forms, location and spatial relationships, and transformations. Because geometry is considered a focal area for preschool and kindergarten, geometry experiences should figure prominently in curriculum planning.

Longitudinal Planning

Block building should be a major focus of the geometry curriculum in both preschool and kindergarten. Through block building, children explore the properties of two- and three-dimensional shapes, develop concepts about location and spatial orientation, and even experience some of the effects of transformations. Long-range planning should provide a wide range of opportunities for children to explore geometric concepts related to both shape and spatial orientation.

Developmental Sequences for Preschool

Children follow predictable sequences in their development of geometric concepts, both in their ability to recognize and label geometric shapes and in their understanding of spatial terms. The goal of long-range planning should be to sequence children's experiences so that they build on their prior knowledge and broaden their conceptual understanding.

Two- and Three-Dimensional Forms   Some children may have had little exposure to geometric forms prior to entering preschool, whereas others may be able to name some shapes. The former would be at a level that precedes the van Hiele (1999) Visual Level of geometric reasoning, whereas the latter would be at the Visual Level. Commensurate with these expected developmental levels, long-range planning should focus on deepening children's ability to recognize shapes in both standard and nonstandard configurations. It should also ensure that children explore shapes throughout the classroom—through art media, science experiments, and a wide variety of manipulative materials.

Developmentally, children gradually progress into van Hiele's Descriptive Level of reasoning, where they begin to use language to describe the properties of shapes. Although preschool children would not be expected to reach this level, experiences should lead in that direction.

Location and Spatial Relations   The preschool years are marked by rapid growth in language. Related to this is children's acquisition of positional terms to identify the location of objects, including themselves. Longitudinal planning should include a regular focus on positional terms through a variety of activities that integrate movement and position with language. Because research documents a progression in children's understanding of these terms, teachers of young preschool children should initially focus on terms related to direct contact with an object, such as "in" and "on." Focus can then shift to proximity, such as "next to" and "beside," followed by terms that require another object as a reference point, such as "behind." See Chapter 5 for specific information related to positional terms.

Transformations   Most preschool children cannot mentally picture the effects of geometric transformations, including turns, flips, and slides. For example, when a shape is rotated, they may no longer believe that it is the same shape. Nevertheless, opportunities to experiment with transformations should be included in long-range planning because they help children focus on the properties of shapes, the stage of geometric reasoning that is on their learning horizon.

Developmental Sequences for Kindergarten

Kindergarten children follow developmental trajectories similar to those described for preschool children, but are usually more advanced in their reasoning. They may begin at a more higher level, particularly if they have had a rich preschool experience. Some children, however, may have had few experiences related to geometry prior to coming to kindergarten.

Two- and Three-Dimensional Forms   Kindergarten children are also likely to be in van Hiele's Visual Level of geometric reasoning, although they may have more knowledge than preschool children. For example, they may know the names of more shapes, and some kindergarten children may be able to identify shapes in nonstandard representations. Like preschool children, kindergartners need many opportunities to explore geometric shapes through various media, including art and manipulative materials. In addition to focusing on shape recognition of both standard and nonstandard forms in various orientations, kindergarten teachers should plan opportunities for children to compose and decompose two- and three-dimensional forms. This helps them focus on the properties of flat and solid shapes.

During kindergarten, some children may begin to transition into van Hiele's second stage, the Descriptive, or Analytic, Level. They may begin to describe the properties of various shapes and group them accordingly. For this reason, a strong emphasis should be placed on describing the properties of shapes at the kindergarten level.

Location and Spatial Orientation   During kindergarten, emphasis on the understanding of positional terms should continue. Children may use more precise terminology when describing location, including ordinal numbers. For example, a child might say that a particular art material is on the third shelf from the bottom. They may also begin to draw graphic representations of locations, such as maps of the classroom or the playground. Teachers should build upon these concepts in their short- and long-term planning.

Transformations   Explorations of geometric transformations should become a focus of the geometry curriculum in kindergarten. Research indicates that by the end of kindergarten, many children are beginning to visualize rotations. Therefore, the kindergarten curriculum should support this conceptual development.

Short-Term Planning

In their short-term planning, teachers should focus on the developmental levels of individual students and plan accordingly. Many materials, such as teacher-made shape games, are self-leveling. Teachers might focus on naming shapes with one child, recognizing shapes in different orientations with another, and describing the properties of shapes with a third.

Planning for Preschool

Geometry manipulative materials should always be available in the preschool classroom. Puzzles, manipulative shapes, and geometric building materials all contribute to children's understanding of geometry concepts. Through individual and small-group work, teachers should plan specifically to focus on composing and decomposing shapes. Art materials and manipulative toys that allow shapes to be combined to create other shapes present a venue for teacher-scaffolded explorations. In addition to supporting children's exploration of geometric manipulative materials, teachers should spend time daily in the block area in order to encourage children's construction of geometric structures. The addition of accessory materials, such as toy people, animals, and vehicles, should be a part of teachers' short-range planning so that both boys and girls are encouraged to use the area.

Periodically, geometry activities should be planned as the special activity of the day. Children can create imprints with geometric shapes dipped into paint or play dough, play shape games, or create class shape books. Teachers should also plan to integrate geometry into at least one other area of the classroom weekly. This could include shape templates or collage materials in the art area, geometry-related books in the literacy area, or an obstacle course in the gross-motor area to emphasize positional terms, such as *over*, *under*, and *through*.

Planning for Kindergarten

Geometry should be one of the focuses of small-group explorations during kindergarten, perhaps during daily math time. Children can work together to compose and decompose shapes, create geometric transformations, play shape games, draw geometric representations, and explore the properties of shapes. Teachers might decide to have children work on number concepts three days a week, algebra once a week, and geometry or measurement once a week.

Geometry can also become a regular part of large-group experiences in kindergarten. Teachers can introduce concepts related to shape, location, and transformations, and read books and sing songs that focus on geometric concepts. Chapter 5 includes ideas for group-time experiences.

PLANNING FOR MEASUREMENT

Measurement concepts are related to number sense, geometry, and data analysis, so planning in those areas can also encompass measurement. In addition, measurement is a component of science and can be integrated into science activities. Two important areas of focus in the Measurement standard are comparing the measurable attributes of objects and using standard and nonstandard measuring tools.

Longitudinal Planning

Long-range planning for teaching measurement should include many opportunities for children to explore measurement within meaningful contexts. This means that as teachers plan for other areas of the classroom, such as science, the sensory table, or the block area, they should consider ways to incorporate various types of measurement. In addition, long-range planning should ensure that children have multiple opportunities to explore concepts of length, weight, volume, area, speed, and distance.

Developmental Sequences for Preschool

Preschool children usually enter school already using some vocabulary to make measurement comparisons. They may also make measurement comparisons on the basis of functionality, such as pairing a big cup with a big plate and a small cup with a small plate. Teachers can help preschool children move from direct comparison of objects to the use of an intermediary object to make comparisons. Older preschool children may begin to develop the concept of using a unit to measure.

Comparing Measurable Attributes   Preschool children judge relative size by making direct comparisons—looking at a big block next to a little block, lifting a bucket filled with sand and an empty bucket, or comparing mounds of play dough. Teachers can begin the year by focusing on direct comparisons in their planning. Materials such as stacking and nesting toys can be planned for the manipulative area, containers of various sizes placed in the water table, and triangles of different sizes displayed in the music area. This type of experience should continue throughout the year so that children have many opportunities to compare the properties of objects in different situations.

While still encouraging direct comparisons, teachers can next introduce the idea of an intermediary object to use to judge size. A string can be used to compare the circumferences of several pumpkins, and a scoop or cup to judge the relative volumes of containers in the sensory table. Over time, teachers can plan materials, for various areas of the classroom, that encourage both direct comparison and comparison through an intermediary object. The Shoe Store activity described in Chapter 6 would be an example.

As the year progresses, some preschoolers may become interested in the idea of a unit for measuring objects. Teachers can spur this interest by introducing books that describe measurement and manipulative materials that encourage it. The interlocking inchworms, described in Chapter 6, are an example.

Using Measurement Tools   The use of measurement tools by preschool children moves from employing random objects to judge size, to more precise measures. At the beginning of the year, when children are asked how tall an object is, they may compare it with a part of their body or find another object to hold next to it. With guided experience, children begin to mark more precise measures on a string, stick, or paper. Throughout the year, teachers can plan measurement experiences that employ a variety of tools. For example, they might give students a straw or a stick to mark the growth changes of a plant, or they might suggest that students use popsicle sticks to measure the lengths of various items in the room. Although most measurement done in preschool employs nonstandard tools, standard measurement tools can also be introduced in order to provide exposure. For example, rulers can be included in the art area. Children may initially use them to draw lines and then notice the unit marks on them.

Chapter 6 recommends the inclusion of a measurement center in the classroom. Such a center could provide children with access to a variety of tools and encourage experimentation with measurement whenever students are interested. Teachers could periodically plan for the introduction of new materials to this center. If measurement received more support and exposure in preschool classrooms, young children might develop measurement concepts more readily than is currently expected.

Developmental Sequences for Kindergarten

Children entering kindergarten are just a few months older than those leaving preschool, so overlap in the measurement curriculum, as well as in other math areas, should be expected. Throughout the year, kindergarten children should experience many opportunities to use units for measuring and should have more exposure to standard measurement tools than during earlier years. Measurement can be included as a regular part of both small- and large-group experiences.

Using Tools to Compare Measurable Attributes   Although kindergarten children continue to use direct comparison and intermediary objects to judge size, they can also begin to use units to measure objects. Individual objects, such as cubes or popsicle sticks, should be used as measurement units so that children can construct the concept of quantifying discrete units when measuring. The units on standard measures, such as rulers, yardsticks, and tape measures, are continuous, and young children are likely to regard the entire tool as one unit. Subdividing a yardstick into feet and inches is a difficult concept that takes a long time for children to understand.

Like preschool teachers, kindergarten teachers may start the year with opportunities for children to use direct measurement comparisons, but move more quickly to the use of an intermediary object. For example, kindergarten children could compare which leaves were larger and smaller than a popsicle stick, and which nuts were larger and smaller than a paper clip. Planning should then begin to focus on unitizing. Children should have many opportunities to use a wide variety of nonstandard units to measure height, circumference, length, and distance. Unitizing can also be incorporated into other types of measurement. For example, children could use a coffee scoop (⅛ cup) to measure the volume of various containers, or pebbles and a balance scale to judge the weight of various objects.

Kindergarten children should also have exposure to standard measuring tools, perhaps in conjunction with nonstandard tools. For example, children could use 1-inch cubes to measure the length of an object and then compare their measure with the marks on a ruler. Some kindergarten classrooms have a container of pencils, glue, and scissors available at each group's table. Rulers could be added to these materials during the course of the year to increase children's exposure to standard measurement tools. A well-stocked measurement center could also add to the measurement opportunities of kindergarten children.

Short-Term Planning

Short-term planning should ensure that measurement experiences are included in weekly plans. Measurement opportunities are underutilized, particularly in preschool. Short-term planning can ensure that measurement concepts are regularly integrated into a variety of areas of the classroom.

Planning for Preschool

Short-term planning for measurement in preschool might begin with brief notes on the lesson plan of areas in which to emphasize measurement. For example, the teacher might make a note to encourage children to compare the lengths and widths of blocks, or to use one size of block to measure their structures. In the sensory table, teachers might plan for children to estimate the relative volumes of different containers and then pour water back and forth among them to make judgments. In both of these examples, the teacher will need to devote some time to the areas in order to support these instructional goals.

Periodically, measurement should be included as a special activity. Students might make hand prints and then compare them with an object such as a crayon, or they might use measurement as part of a cooking activity. Teachers should also consider ways to integrate measurement into other areas of the classroom, as part of their regular short-term planning. This might involve including a scale and tape measure in a dramatic play doctor's office, or measuring cups in the sensory table. These are easy adjustments to make to the classroom environment; however, if teachers do not consciously plan for them, the opportunities may be lost.

Planning for Kindergarten

As with algebra and geometry, measurement can be added to small-group math experiences in kindergarten. If the class has a daily math period, then number sense might be a focus for three days of the week, whereas measurement, algebra, and geometry alternate on the other days. Because number sense is related to all of these areas, they can be used to support emerging number-sense concepts while also introducing children to new mathematical ideas.

In addition to including measurement activities regularly during short-term planning, teachers should also note areas of the classroom or other areas of the curriculum in which measurement can be integrated. For example, if children are using ramps for physical-knowledge experiments, then measurement would be a logical contributor to that process. When children are reading and discussing animals, then finding items in the classroom that are of an equivalent length to the various animals would add to their comprehension.

PLANNING FOR DATA ANALYSIS AND PROBABILITY

The Data Analysis standard supports the understanding of algebra and number concepts. Because sorting objects by a specific attribute is directly related to algebra, when teachers plan activities that encourage children to sort and group data, they are accommodating both the Algebra and Data Analysis standards. In addition, when the data are organized and analyzed, as on a group graph, the Number and Operations standard comes into play. The graph allows children to quantify and compare groups more easily, thereby supporting number-sense concepts while also introducing children to the use of data in mathematics. Estimation and probability experiences can contribute to number sense if they are not too far beyond children's level of understanding, because they encourage children to think about number relationships in a new way.

Longitudinal Planning

Data analysis and probability activities should be planned periodically to introduce new concepts, such as likelihood, and to support developing concepts in other mathematical areas, such as comparison of sets. Graphing, in particular, is of interest to both preschool and kindergarten children when they vote on classroom situations. For this reason, graphing experiences should be included throughout the year. Estimation and probability activities may be included periodically in long-range planning so that children have exposure to these concepts.

Developmental Sequences for Preschool

Developmental sequences are less well researched in the areas of data analysis and probability than in other mathematical areas. Information from teachers who have incorporated experiences in these areas into their classrooms, however, indicates that both preschool and kindergarten children can engage productively in some activities in these areas if the activities are directly related to concrete experiences.

Data Analysis   Data analysis begins with the sorting of data into groups. The topic of sorting and classifying objects according to their attributes has already been discussed in the algebra section and will therefore not be a focus of this section. The display of data on bar graphs, however, extends the likelihood that children will analyze the data by making comparisons among groups; therefore, it is a focus of this standard. The introduction of group graphs is probably most appropriate in older preschool classrooms; however, younger or less advanced children also eagerly participate. Graphing should be included regularly throughout the school year because it strongly supports number-sense concepts. Children at all levels of quantification can make magnitude judgments by looking at class bar graphs. In addition, the questioning that accompanies graphing activities involves mathematical language that is important for children to understand. Finally, graphing is a way for teachers to more concretely represent voting responses of children. For these reasons, teachers should include group graphing at least monthly in their longitudinal planning.

Estimation and Probability   Estimation activities stretch children's number-sense concepts, because children must use some existing data to envision possible totals. Because young children view estimation activities as games, they enjoy participating when small quantities that are within their scope of reasoning are included. Estimation activities should be planned periodically, particularly for older preschool children.

Probability is a difficult concept for young children and not a focus of the preschool curriculum. Inclusion of beginning concepts, however, such as "likely or unlikely," can be incorporated occasionally in the teacher's short-term planning to provide children with opportunities to think about and discuss the concepts.

Developmental Sequences for Kindergarten

Kindergarten children can explore concepts related to Data Analysis and Probability in more depth. Specifically, they can move from a whole-group introduction of the concepts to working together in small groups and, eventually, completing individual activities. For example, kindergarten children are much more capable of constructing individual bar graphs than preschool children, particularly after they have had many experiences with group graphs.

Data Analysis   Like preschool children, kindergartners like to visualize the results of their voting experiences on class bar graphs. They can also analyze other types of data displayed on graphs, such as the number of each type of animal that the class saw on a field trip. For the first part of the year, graphs should be constructed at the large-group level. This gives children experience with graphing concepts and allows teachers to provide necessary support.

As the year progresses, teachers may introduce graphing at the small-group level. Children can work together to graph collections of objects, perhaps during small-group math sessions. By the end of the year, teachers can incorporate individual graphing into their planning. Although not all children may be able to accurately represent data on graphs, the experiences give teachers assessment data to use as they scaffold children's understanding of graphing concepts.

Estimation and Probability   Because kindergarten children, in general, have more highly developed number-sense concepts than preschool children, they can engage more successfully in estimation activities. At first, small quantities should be used. Depending on children's success at this level, teachers can gradually introduce larger quantities into estimation experiences. The questions and discussions that accompany estimation activities, such as which estimates were closest to the exact amount, support construction of a mental number line.

Probability activities, such as deciding which color of block is more likely to be pulled from a bag, are of interest to kindergarten children, although they may not yet be logical in their reasoning. At first, the items in the activity should be visible to the children throughout the experience. Later, teachers can plan similar activities in which children see the objects as they are placed in the container, but cannot see them thereafter. Probability activities should be introduced periodically to engage children in thinking about possibilities; however, they are not a major focus of the curriculum.

Short-Term Planning

Short-term planning allows teachers to coordinate data analysis and probability activities with other topics that are of interest in the classroom. For example, if teachers know that their class will be voting on a name for a new class pet, then inclusion of a graphing activity to display the results would be a logical extension. If children are interested in small nuts and pine cones that have been collected during autumn, then estimating how many of each will fit into a container would connect this interest to mathematics. Short-term planning helps make mathematical concepts relevant to the lives of children.

Planning for Preschool

Graphing activities in preschool often go along with voting experiences during group time. The graphs provide a venue for teachers to model counting and scaffold children's comparison of groups. These are important concepts in preschool. For this reason, teachers should remember to include group graphing in their short-term plans. The graphs should be left on display so that children can return to them from time to time to discuss the results.

Periodically, estimation should also be included in short-term planning. This could be in conjunction with a science display or another topic of interest in the classroom. For example, in a class that was interested in babies, children estimated the number of pacifiers in a clear jar.

Planning for Kindergarten

Graphing and estimation should be included regularly in kindergarten because they reinforce and extend number-sense concepts. To add to the small-group model that has emerged throughout this chapter, graphing and estimation could be incorporated into regular math sessions. With number-sense activities planned for three days a week, graphing and estimation could alternate with algebra, geometry, and measurement on the remaining days. This would ensure regular exposure to data analysis and estimation concepts.

Graphing, estimation, and probability can also be included regularly in group-time planning. The concepts might be introduced to the large group and incorporated at the small-group level. For example, at group time the teacher might introduce a jar containing four black marbles and one white marble. During small-group math sessions, children could take turns drawing a marble from the jar and recording the results. They could report their conclusions at the next large-group session.

Assessment

Assessment strategies should be in place before the curriculum is implemented. Otherwise, teachers may forget to collect assessment data during the activities. Preschool and kindergarten classrooms are busy places. If a checklist, anecdotal notebook, or digital camera is not readily at hand, important information may be forgotten. Ongoing assessment information is necessary for teachers to plan effectively for individuals and the class as a whole.

Questioning strategies are an important part of assessment. Because teachers must adjust their responses according to children's reactions to activities, specific questions cannot be planned ahead of time. However, teachers can list the types of questions that they plan to ask in order to gain assessment information. The list can be carried in a pocket or notebook for quick access.

SAMPLE SCHEDULES

At first, developing a comprehensive math curriculum may seem like an overwhelming undertaking. There are math domains to consider, multiple topics within those domains, and varying levels of development for each topic. The sample planning schedules that follow are included to provide graphic models for long- and short-term planning (see Tables 8.1–8.4). They are examples that teachers can adapt to meet their own needs. These schedules are formulated on the basis of general developmental expectations for preschool and kindergarten classes. As always, teachers will need to make adaptations for individual students and, perhaps, for the class as a whole. For example, path games, which are not included on the sample preschool plan during the first months of school, may be appropriate for some preschool classes from the beginning of the year. However, some kindergarten classes may start at a lower level than that which is represented on the sample kindergarten plan. Teachers should always have a variety of dice available to quickly adapt games to the developmental levels of the students. In most cases, general curriculum activities are noted on the sample longitudinal plans; however, in some cases, specific materials or activities are listed because they are particularly appropriate for an initial activity or for a particular time of the year.

The tables indicate either the introduction of new types of materials and concepts or an area of focus. It is expected that materials and activities which support concepts highlighted earlier on the schedule will continue to be available, as all of these big ideas in math take a long time to develop. As concepts are extended throughout the year, teachers should develop new materials to support them. For example, collections should be available for sorting and classifying throughout the year in preschool, but a variety of materials should be used for these collections. Very few children will be interested in sorting bottle caps all year long. It is also important to remember that kindergarten children will experience the same types of activities at a higher, expanded, or more in-depth level than preschool children will.

The tables charting short-term planning for preschool and kindergarten correlate with a selected month from the analogous long-term planning table. Many more activities are included on the short-range plans. This is because focus topics on the longitudinal plan can be targeted in many different ways. The short-term plans also represent carryover activities from the previous month, as well as mathematical concepts that are included for exposure, but are not a main topic on the long-range plan. Note that these lesson plans reflect only math-related activities, not all of the activities that would be planned for the class.

Table 8.1. Long-term scheduling guide for preschool

	September	October	November	December	January
Number sense	Manipulative games, 1–3 dot die	Manipulative games, 1–3 or 1–6 dot die	Manipulative games, 1–3 or 1–6 dot die	Manipulative games, 1–6 dot die	Manipulative games, 1–6 or 1–10 dot die
	Simple grid games, 1–3 dot die	Simple grid games, 1–3 or 1–6 dot die	Grid games, 1–3 or 1–6 dot die	Grid games, 1–3 or 1–6 dot die	Grid games, 1–6 dot die
	Counting songs to 5	Counting songs, forward and backward, to 5	Short path games, 1–3 dot die	Short path games, 1–3 dot die	Short path game, 1–3 or 1–6 dot die
	1:1 correspondence focus throughout	Focus on comparison of small sets	Counting songs to 10	Counting songs to 10	Long path games, 1–6 dot die
		Counting books	Focus on comparison of small sets	Focus on comparison of small sets	Focus on addition or subtraction of 1
			Model numerals for cardinal sets	Model numerals for cardinal sets	Scoring in motor room
					Model numerals for cardinal sets
Algebra	Bottle caps for sorting	Fall collections to sort	Collection to sort	Collection to sort	Chanting visual patterns
	Clap name patterns	Movement patterns	Movement patterns	Rhythm patterns	Creating math books or journals
			Patterns in songs	Reenactment of math in story	
Geometry	Shape manipulatives	Shape games	Shapes in art	Shape games, standard and nonstandard forms	Focus on blocks
	Blocks	Blocks	Blocks	Identifying shapes in the environment	Composing and decomposing shapes
		Focus on location words	Location and movement words	Combining shapes	Specifying location
Measurement	Nesting materials	Materials to seriate	Measure, direct comparison and intermediate object	Focus on volume	Measurement storybook
	Two sizes of wood blocks for music	Measurement in science (pumpkins or gourds)		Direct and indirect comparisons	Introduce unit measure
Data analysis and probability	Group graphing	Group graphing	Group graphing	Group graphing	Group graphing
		Estimation, 3–5 items	Estimation, 3–5 items	Estimation, 4–6 items	Estimation, 5–7 items

(continued)

Table 8.1. (*continued*)

	February	March	April	May	June End of school year
Number sense	Manipulative games, 1–6 or 1–10 dot die Grid games, 1–6 dot die Short path games, 1–6 dot die Long path games, one or two 1–6 dot dice Focus on addition or subtraction of 1	Manipulative games, 1–6 or 1–10 dot dice Grid games, 1–6 dot die Short path games, 1–6 dot die Long path game, one or two 1–6 dot dice Composing and decomposing small numbers	Manipulative games, 1–6 or 1–10 dot die Grid games, 1–6 dot die Short path games, 1–6 dot die Long path game, one or two 1–6 dot dice Composing and decomposing small numbers	Manipulative games, 1–6 or 1–10 dot die Grid games, 1–6 dot die Short path games, 1–6 dot die Long path game, one or two 1–6 dot dice Composing and decomposing small numbers	Assemble and distribute children's progress reports and portfolios.
Algebra	Patterning materials Chanting patterns from class materials	Patterning activities throughout classroom	Patterning activities throughout classroom	Patterning activities throughout classroom	
Geometry	Comparison of two- and three-dimensional forms Focus on three-dimensional forms in science	Shape games, two- and three-dimensional forms Specifying location	Shape games, two- and three-dimensional forms Composing and decomposing shapes Specifying location	Shape games, two- and three-dimensional forms Composing and decomposing shapes Specifying location	
Measurement	Continue unit measures Exposure to standard measures	Focus on area Introduce measurement center	Measurement throughout classroom Measurement center	Measurement throughout classroom Measurement center	
Data analysis and probability	Graphing collections, individual and small groups Group graphing Estimation, 5–8 items *Likely, unlikely* discussions	Graphing collections, individual and small groups Group graphing Estimation, 5–8 items *Likely, Unlikely* discussions	Graphing collections, individual and small groups Group graphing Estimation, 5–10 items Simple probability activities	Graphing collections, individual and small groups Group graphing Estimation, 5–10 items Simple probability activities	

Table 8.2. Long-term scheduling guide for kindergarten

	September	October	November	December	January
Number sense	Manipulative games, 1–6 dot die	Manipulative games, 1–6 or 1–10 dot die	Manipulative games, 1–6 or 1–10 dot die	Card games, set comparisons	Card games, set comparisons, and addition of small sets
	Grid games, 1–6 dot die	Grid games, 1–6 or 1–10 dot die	Grid games, 1–6 or 1–10 dot die	Grid and path games, adding two 1–3 dot dice	Path games, adding two 1–3 or 1–6 dot dice
	Short path games, 1–3 or 1–6 dot die	Short path games, 1–6 dot die	Short path games, one 1–6 or two 1–3 dot dice	Composing and decomposing numbers to 6	Composing and decomposing numbers to 6
	Counting songs to 5	Long path games, 1–6 dot die	Long path games, 1–6 or 1–10 dot die	Scoring games	Scoring games
	Counting focus throughout classroom	Counting songs and books to 10	Counting books to 10 and beyond		
		Counting focus, forward and backward	Composing and decomposing numbers to 5		
		Modeling and use of numerals when writing cardinal quantities	Modeling and use of numerals when writing cardinal quantities		
		Use of ordinal numbers			
Algebra	Collection for sorting	Collection for sorting	Chant simple visual patterns	Patterning throughout the curriculum	Sorting objects by two simultaneous criteria
	Clap name patterns	Chant simple visual patterns	Individual and small-group patterning with objects	Chanting more complex patterns	Patterns in nature
	Rhythm patterns	Rhythm and movement patterns		Recognizing patterns in many forms	Extending more complex patterns
	Movement patterns				Recognizing patterns in many forms
Geometry	Shape manipulatives	Shape games, nonstandard forms and orientation	Shape games, two- and three-dimensional forms	Constructing and deconstructing shapes	Experimenting with three-dimensional shapes
	Shape games	Blocks	Rotations	Rotations and slides	Flips
	Blocks	Focus on location words	Specifying location	Specifying location	Simple mapping
Measurement	Seriation activities	Seriation activities	Direct comparison and intermediate object	Unitizing	Comparing various units
	Three sizes of wood blocks, music	Measurement in science (pumpkins or gourds)	Volume measurement	Measurement center	Measurement center
				Volume measurement	
Data analysis and probability	Group graphing	Group graphing	Group graphing	Group graphing	Group graphing
	Estimation, 3–6 items	Estimation, 4–8 items	Individual and small-group graphing	Likely and unlikely; possible and impossible	Necessary and not necessary
			Estimation, 5–10 items		

(continued)

Table 8.2. *(continued)*

	February	March	April	May	June End of school year
Number sense	Card games, set comparisons and addition of small sets	Card games, set comparisons and addition of small sets	Card games, set comparisons, and addition of medium sets	Card games, set comparisons, and addition of medium sets	Assemble and distribute children's progress reports and portfolios.
	Path games, adding two 1–3 or 1–6 dot dice	Path games, adding two 1–6 dot or numeral dice	Path games, adding two 1–6 dot or numeral dice	Path games, adding two 1–6 dot or numeral dice	
	Addition and subtraction manipulative games	Addition and subtraction manipulative games	Addition and subtraction manipulative games	Addition and subtraction manipulative games	
	Composing and decomposing numbers to 6	Composing and decomposing numbers to 10	Composing and decomposing numbers to 10	Composing and decomposing numbers to 10	
	Scoring games	Scoring games	Scoring games	Scoring games	
Algebra	Sorting objects by two simultaneous criteria	Sorting objects with overlapping criteria (intersecting sets)	Sorting objects with overlapping criteria (intersecting sets)	Representing the same pattern in multiple ways	
	Representing the same pattern in two ways (red–blue and clap–snap)	Representing the same pattern in two ways	Representing the same pattern in three ways		
Geometry	Constructing and deconstructing two- and three-dimensional forms	Constructing and deconstructing two- and three-dimensional forms	Geometry problems in stories	Rotations, slides, and flips	
	Rotations, slides, and flips	Rotations, slides, and flips	Lines, parallel and intersecting	Lines	
			Symmetries	Symmetries	
Measurement	Unit measures	Unit measures	Measuring objects in many different ways	Measuring objects in many different ways	
	Comparison to standard measures	Comparison to standard measures	Measurement center	Measurement center	
	Area measurement	Weight measurement			
Data analysis and probability	Group graphing	Group graphing	Group graphing	Group graphing	
	Individual and small-group graphing	Individual and small-group graphing	Individual and small-group graphing	Individual and small-group graphing	
	Estimation, 6–12 items	Simple probability activities	Simple probability activities	Simple probability activities	

Table 8.3. Short-term scheduling guide for preschool mathematics: month of October, from Table 8.1

WEEK 1 ACTIVITIES	Monday	Tuesday	Wednesday	Thursday	Friday
Large group	Name song (1:1 correspondence) Five Little Leaves finger-play (counting; see Figure 8.1) Clap children's names (patterning)	Name song (1:1 correspondence) Five Little Leaves finger-play (counting; see Figure 8.1) Clap children's names (patterning)	Story—*Pumpkin, Pumpkin,* (Titherington, 1990) Reread, with children showing the size of the plant (measurement)	Estimate pumpkin counters in a clear jar (number sense, probability)	Five Little Leaves finger-play (counting; see Figure 8.1) Movement—fall slowly like a leaf, and fast like a nut (measurement)
Individual/ small Group (Monday–Friday choice time)	Pumpkin manipulative game, pumpkin counters and 1–4 spinner (quantification and set comparison) Leaf grid game, 1–6 dot and marble chip counters (quantification and set comparison) Collection of nuts for sorting and classifying (algebra and data analysis) Shape sorter, six standard shapes (geometry) Shape bingo game, standard and nonstandard orientations (geometry) Interlocking shapes for building (geometry, measurement)				
Special	Orange playdough, three sizes of pumpkin cookie cutters (measurement)			Shape bingo game (geometry)	
Integrated (Monday–Friday choice time)	Seriate pine cones in science area (measurement) Farmer's market in dramatic play, with fruits and vegetables to sort (algebra and data analysis) Dried corn in sensory table with various sizes of containers, scoops (measurement) Fall counting books				

WEEK 2 ACTIVITIES	Monday	Tuesday	Wednesday	Thursday	Friday
Large group	Five Little Pumpkins finger-play (counting; see Figure 8.1) Book—*Count Down to Fall* (Hawk, 2009)	Big-little pattern with pumpkins—children raise and lower hands to indicate pattern (algebra)	Five Little Pumpkins finger-play (counting; see Figure 8.1) Graph predictions about how pumpkins grow	Field trip—pumpkin farm	Compare photos from field trip to class graph
Individual/ small group (Monday–Friday choice time)	Pumpkin manipulative game, pumpkin counters and 1–4 spinner Leaf grid game, 1–6 dot and marble chip counters Collection of nuts for sorting and classifying Shape sorter, six standard shapes Shape bingo game, standard and nonstandard orientations Interlocking shapes for building				

(continued)

185

Table 8.3. (continued)

WEEK 2 ACTIVITIES	Monday	Tuesday	Wednesday	Thursday	Friday
Special		Painting with two sizes of pumpkin cookie cutters (measurement)			Field trip books—children sequence trip (ordinal numbers)
Integrated (Monday–Friday choice time)	Various sizes of pumpkins and gourds in the sensory table, along with various sizes of tongs Farmer's market in dramatic play, with fruits and vegetables to sort Three sizes of triangles in music area (measurement) Fall counting books				

WEEK 3 ACTIVITIES	Monday	Tuesday	Wednesday	Thursday	Friday
Large group	Five Little Pumpkins finger-play (counting; see Figure 8.1) Seriate five pumpkins from field trip (measurement)	Apple Song Game (Chapter 3) (counting, subtraction) Book—From the Garden: Counting Book (Dahl, 2004)	Five Little Pumpkins finger-play (counting; see Figure 8.1) Graph children's votes on what to cook with pumpkins (data analysis, comparison of sets)	Apple Song Game (Chapter 3) (counting, subtraction) Estimate buckeyes in a jar (number sense, probability)	Apple Song Game (Chapter 3) (counting, subtraction) Chant patterns with apples and pumpkins (algebra)
Individual and small group (Monday–Friday choice time)	Hi Ho! Cherry-O, commercially available game (quantification, subtraction, and set comparison) Pumpkin grid game, 1–6 dot die and black cat counters (quantification and set comparison) Collection of novelty pumpkins and gourds to sort (algebra and data analysis) Shape attribute blocks (geometry) Pattern blocks (geometry) Magnetic shapes for building (geometry, measurement)				
Special		Art—shape imprints in clay			Cooking—making pumpkin bread (measurement)
Integrated (Monday–Friday choice time)	Science—comparing sizes of pumpkins with various sizes of brass hoops (measurement) Shape templates in art area (geometry) Pumpkin seeds in sensory table to count and put in small containers (number sense) Farmer's market in dramatic play, with price tags and pennies (number sense) Fall counting books				

WEEK 4 ACTIVITIES	Monday	Tuesday	Wednesday	Thursday	Friday
Large group	Apple Song Game (Chapter 3) (counting, subtraction) Chanting apple color patterns	Counting book—*One Too Many* (Marino, 2010) Chanting apple color patterns	Apple Song Game (Chapter 3) (counting, subtraction) Graphing—favorite kind of apple	Sort fruits by shape—apple, orange, lemon, lime, grape, strawberry, papaya, starfruit, watermelon, cantaloupe, peach, pear (algebra, geometry)	Counting book—*One Too Many* (Marino, 2010)—search for specific clues (algebra)
Individual and small group (Monday–Friday choice time)	Hi Ho! Cherry-O, commercially available game (quantification, subtraction, and set comparison) Pumpkin grid game, 1–6 dot die and black cat counters (quantification and set comparison) Collection of novelty pumpkins and gourds to sort (algebra and data analysis) Shape attribute blocks (geometry) Pattern blocks (geometry) Magnetic shapes for building (geometry, measurement)				
Special	Art—creating signs with shape templates (geometry)		Cooking applesauce—measuring apple peels and ingredients		
Integrated (Monday–Friday choice time)	Science—weighing fresh and dried gourds Shape templates in art area (geometry) Farmer's market in dramatic play, with price tags and pennies (number sense) Buckeyes in sensory table, with clear containers of various sizes (counting, estimation) Fall counting books				

Table 8.4. Short-term scheduling guide for kindergarten month of March, from Table 8.2

WEEK 1 ACTIVITIES	Monday	Tuesday	Wednesday	Thursday	Friday
Large group	Sort flowers that have solid and mixed colors into overlapping circles (algebra)	Flower patterns in sun bonnet (algebra)	Plant three types of seeds—insert straws for measuring	Weekly Geometry (Chapter 5)—transformations	*Each Orange Had Eight Slices* (Giganti, 1999) (number sense)
Individual work or small math groups	Small groups repeat group sorting activity (algebra)	Flower long-path collection game—adding two dice (number and operations)	I Have More addition card game (Chapter 2), played in pairs (number and operations)	Children sort bags of seeds and graph them (algebra, data analysis)	Flower long-path collection game—adding two dice Represent total of each flower collected in math journal (number and operations)
Special			Children plant 3 seeds and insert straws		
Integrated or choice time	Flower shop in dramatic play, with price tags, play money, flowers to sort (number sense and algebra) Artificial flowers for creating patterns (algebra); seed collection to sort on graph board (algebra, data analysis) Block building Flowers to sort Geometry pattern blocks				

WEEK 2 ACTIVITIES	Monday	Tuesday	Wednesday	Thursday	Friday
Large group	*Each Orange Had Eight Slices* (Giganti, 1999) include problems from book	Flower patterns in sun bonnet (algebra)—students represent by chanting	Compare seed growth; model marking growth on straws	Weekly Geometry (Chapter 5)—composing shapes	*Each Orange Had Eight Slices* (Giganti, 1999) include problems from book
Individual work or small math groups	Flower long-path collection game—adding two dice (number and operations)	Create pictures by combining precut geometric shapes	Children measure seed growth and mark height on straws (measurement)	I Have More addition card game, played in pairs (number and operations)	Group choice of class math games (number and operations)
Special	Fold-over paintings of flowers to create flips (geometry)				
Integrated or choice time	Same as Week 1				

WEEK 3 ACTIVITIES	Monday	Tuesday	Wednesday	Thursday	Friday
Large group	*Eggs and Legs: Counting by Twos* (Dahl, 2005) (number sense)	Flower patterns in sun bonnet—students represent with cubes (algebra)	Compare seed growth; use cubes to measure height of growth marks	Weekly Geometry (Chapter 5)—transformations	Group graph—favorite type of flower (data analysis)
Individual work or small math groups	Subtraction card game—counters available	Flower long-path collection game—adding two dice (number and operations)	Children mark seed growth and use cubes to compare heights (measurement)	Subtraction card game—counters available (number and operations)	Create individual flower books—ways to make five
Special	Children compare weights of different flower pots and add marbles to balance				
Integrated or choice time	*Add:* Scales, flower pots, and marbles to choice centers (measurement)				

WEEK 4 ACTIVITIES	Monday	Tuesday	Wednesday	Thursday	Friday
Large group	Show Styrofoam solids that have been sliced—children identify two-dimensional faces	Flower patterns in sun bonnet (algebra)—students represent with movements	Compare seed growth; use cubes to measure height; compare with ruler and yardstick	Weekly Geometry (Chapter 5)—decomposing shapes	Group flowers in pairs and count both individual flowers and pairs
Individual work or small math groups	I Have More addition card game, played in pairs (number and operations)	Math journals—students create a pattern and represent it in two ways	Children mark seed growth, use cubes to compare heights, and compare with ruler (measurement)	Subtraction card game—counters available (number and operations)	Math journals—show three ways to make the number 5 (number sense, algebra)
Special	Children dip sliced faces of solids into paint and create imprints (geometry)				
Integrated or choice time	*Add:* Geometric solids cut into sections for exploration				

REFERENCES

Copple, C., & Bredekamp, S. (2009). *Developmentally appropriate practice in early childhood programs* (3rd ed.). Washington, DC: National Association for the Education of Young Children.

Gersten, R., & Chard, D. (1999). Number sense: Rethinking arithmetic instruction for students with mathematical disabilities. *The Journal of Special Education, 33*(1), 18–28.

Dahl, M. (2005). *Eggs and legs: Counting by twos.* Mankato, MN: Picture Window Books.

Dahl, M. (2004). *From the garden: A counting book about growing food.* Mankato, MN: Picture Window Books.

Giganti, P. (1999). *Each orange had eight slices.* New York: Greenwillow.

Hawk, F. (2009). *Count down to fall.* Mt. Pleasant, SC: Sylvan Dell Publishing.

Hiebert, J., & Carpenter, T.P. (1992). Learning and teaching with understanding. In Grouws, D.A. (Ed.), *Handbook of teaching on mathematics teaching and learning* (p. 65). Reston, VA: National Council of Teachers of Mathematics.

Hutchins, P. (1986). *1 hunter.* New York: Greenwillow Books.

Kato, Y., Kamii, C., Ozaki, K., & Nagahiro, M. (2002). Young children's representations of groups of objects: The relationship between abstraction and representation. *Journal for Reasearch in Mathematics Education, 33*(1), 30–46.

Klibanoff, R.S., Huttonlocher, J., Vasilyeva, M., & Hedges, L.V. (2006). Preschool children's mathematical knowledge: The effect of teacher 'math talk.' *Developmental Psychology, 24*(1), 59–69.

Marino, G. (2010). *One too many.* San Francisco: Chronicle Books.

National Council of Teachers of Mathematics. (2000). *Curriculum and evaluation standards for school mathematics.* Reston, VA: Author.

National Council of Teachers of Mathematics. (2006). *Curriculum focal points.* Reston, VA: Author.

Pinczes, E.J. (1999). *One hundred hungry ants.* New York: Houghton Mifflin.

Titherington, J. (1990). *Pumpkin pumpkin.* New York: Greenwillow.

van Hiele, P.M. (1999, February). Developing geometric thinking through activities that begin with play. *Teaching Children Mathematics, 6*, 310–16.

Williams, S. (1992). *I went walking.* New York: Harcourt.

Glossary

abstraction   One of five principles in counting; the understanding that abstract entities, such as ideas, can be quantified in the same manner as concrete items.

cardinality   The concept that, in counting, the number word assigned to the last item counted in a group of items represents the number of items in the group.

conservation   In child development, the ability to logically reason that a physically transformed object retains the same amount of material.

continuous quantity   An amount that can be infinitely divided into smaller parts, such as clay in a bowl.

count all   An addition strategy in which children count each addend separately before counting all of the items together.

count on   An addition strategy in which children count forward from one of the addends to determine the sum.

counting   The means of quantification that children select when they understand cardinality; one of the three stages of quantification described by Piaget.

discrete quantity   An amount that can be divided into units, such as a set of apples.

disequilibrium   Experiences that conflict with what is already known or predicted to happen and therefore create cognitive conflict.

dynamic image   An image that can be mentally moved into another position or orientation.

equilateral   In geometry, said of a polygon in which all the sides are the same length (see the appendix for additional definitions of polygons).

equilibrium   The incorporation of new experiences into existing cognitive structures that creates a feeling of cognitive balance between old and new ideas.

estimation   A logical approximation when not all the data are available.

face   In geometry, one of the polygons that make up the boundaries of a geometric solid.

flip   A geometric transformation that shows the mirror image of an object; also called a reflection.

formative assessment   An ongoing evaluation that provides both teachers and students with information to support learning.

function   In mathematics, the idea that one quantity (called the input or argument) can determine another quantity (called the output or value).

global quantification   The use of perceptual clues to determine quantity; one of the three stages of quantification described by Piaget.

horizontal axis   On a graph, a line that indicates positions to the left or right of a central point; also called the *x*-axis.

inclusive education   The kind of education in which students with identified disabilities are placed in classrooms together with students without diabilities.

isosceles   In geometry, a type of polygon in which two of the sides are equal, such as an isosceles triangle.

kite   In geometry, a four-sided figure in which two sets of adjacent lines on either side of a line of symmetry are the same length; also called a deltoid.

linear measurement   Measurement along a straight line.

logical–mathematical knowledge   In Piagetian theory, the type of knowledge that is constructed internally as children reflect upon their interactions with objects and form important relationships, such as total quantity, more, less, and the same.

one-to-one correspondence   A relationship in which each item in a group is paired with exactly one item from a corresponding group; in counting, the understanding that one number word must be paired with one (and only one) item being counted; one of the three stages of quantification described by Piaget.

order irrelevant   One of five principles in counting; the understanding that the order in which items are counted does not affect the total.

ordinal number   Number that indicates the position of order of an object, such as first, second, and so forth.

ostinati   In music, rhythmic or melodic patterns that persistently repeat; singular, *ostinato*.

oval   Not well defined in geometry; a type of curve that resembles an egg shape but is not an ellipse.

pattern   A relationship involving a repeating element.

physical knowledge   In Piagetian theory, the type of learning that involves the physical properties of objects, such as color, texture, weight, and shape, and is acquired through directly exploring the objects.

quantification   Determination of the magnitude of a set of items.

reflection   A geometric transformation that shows the mirror image of an object; also called a flip.

reflective symmetry   Similarity of form on either side of a dividing line, called the axis, in which one side is the mirror image of the other; also known as bilateral symmetry.

rotation   A geometric transformation in which an object is repositioned by turning it a particular number of degrees; also called a turn.

rotational symmetry   Similarity of form when an object is rotated (turned) a particular number of degrees around a point; for example, a square is rotationally symmetrical because, when it is rotated 90°, it looks the same as when it is in its original position.

self-leveling   Type of material or activity that can be used effectively by children at various levels of development.

seriate   To order objects by some attribute, such as length.

short-cut sum   An addition strategy in which children count all of the items together to get the sum without first counting the individual addends.

sign   In Piagetian terminology, the representation of an item through an arbitrary mark, such as a numeral, that does not resemble what it represents (in the case of a numeral, a number).

slide   A geometric transformation in which every point on an object moves the same distance and direction; also called a translation.

social knowledge   In Piagetian theory, the type of knowledge that is arbitrary, such as labels for objects, rules, and customs, and is learned through direct transmission of the information from a more knowledgeable person.

spatial orientation   An understanding of the body's position in relationship to physical space, together with the ability to navigate through that environment.

spatial reasoning   The ability to visualize and mentally manipulate spatial patterns.

spatial visualization   The ability to mentally produce and manipulate visual images of two- and three-dimensional objects.

stable order   One of five principles in counting; the understanding that number words must be sequenced in an invariant order.

static images   Images that can be recalled, but not mentally moved into another position or orientation.

subitize   To immediately recognize the number of items in a group, without needing to count them.

symbol   In Piagetian terminology, representation of an object through something that resembles the actual object, such as a picture of it.

symmetry   Similarity of form, arrangement, or design on either side of a dividing line or around a point.

transformation   A repositioning of a geometric shape, through a flip, turn, or slide, without changing the dimensions of the object (i.e., the transformed object is *congruent* with the original); or a proportional change in a geometric shape, in which the angles remain the same but the transformed object is proportionally larger or smaller than the original object (i.e., the transformed object is *similar* to the original).

translation   A geometric transformation in which every point on the object moves the same distance and direction; also called a slide.

turn   A geometric transformation in which an object is repositioned by rotating it a particular number of degrees; also called a rotation.

unitize   In measurement, to use identical, discrete objects to determine size.

universal design for learning (UDL)   Principles for teaching inclusive populations; the principles include multiple means of representation, expression, and engagement.

zone of proximal development (ZPD)   In Vygotskian theory, a distance between mental development and potential development which is close enough that learning from a more knowledgeable person is effective.

Appendix

Identifying Geometric Figures

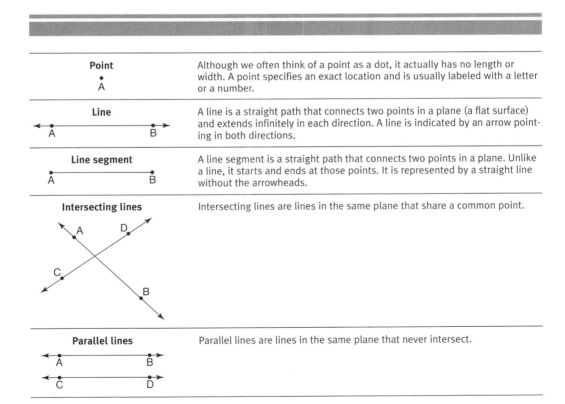

Point • A	Although we often think of a point as a dot, it actually has no length or width. A point specifies an exact location and is usually labeled with a letter or a number.
Line A — B	A line is a straight path that connects two points in a plane (a flat surface) and extends infinitely in each direction. A line is indicated by an arrow pointing in both directions.
Line segment A — B	A line segment is a straight path that connects two points in a plane. Unlike a line, it starts and ends at those points. It is represented by a straight line without the arrowheads.
Intersecting lines	Intersecting lines are lines in the same plane that share a common point.
Parallel lines	Parallel lines are lines in the same plane that never intersect.

Polygon

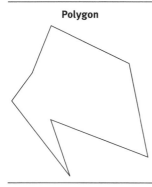

A polygon is a figure formed when three or more line segments, all of which are on the same plane, are joined, and the endpoints of each line segment intersect the endpoints of exactly two other line segments. This figure is an irregular polygon because the length of the sides and the sizes of the angles vary.

Regular polygon

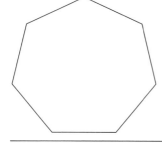

A regular polygon is a polygon in which all of the sides are the same length and all of the angles are the same number of degrees.

Triangle

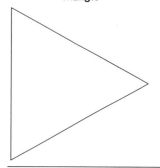

A triangle is a polygon with three sides.

Quadrilateral

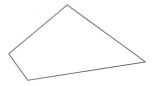

A quadrilateral is a polygon with four sides.

Parallelogram

A parallelogram is a quadrilateral in which both pairs of opposite sides are parallel.

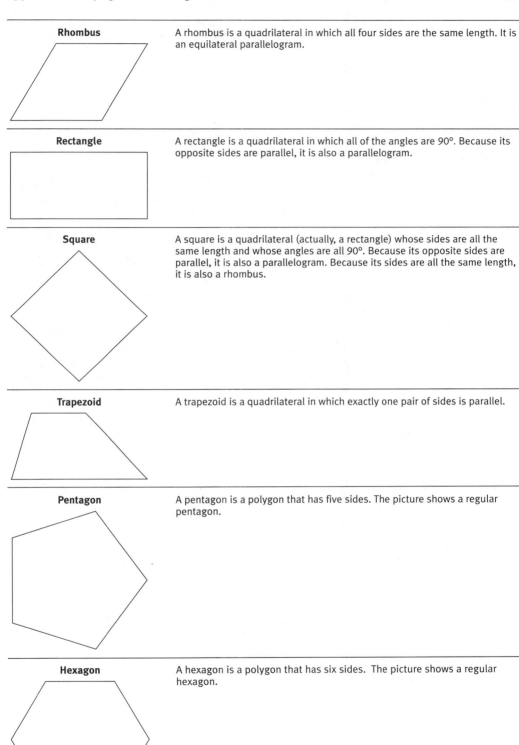

Rhombus

A rhombus is a quadrilateral in which all four sides are the same length. It is an equilateral parallelogram.

Rectangle

A rectangle is a quadrilateral in which all of the angles are 90°. Because its opposite sides are parallel, it is also a parallelogram.

Square

A square is a quadrilateral (actually, a rectangle) whose sides are all the same length and whose angles are all 90°. Because its opposite sides are parallel, it is also a parallelogram. Because its sides are all the same length, it is also a rhombus.

Trapezoid

A trapezoid is a quadrilateral in which exactly one pair of sides is parallel.

Pentagon

A pentagon is a polygon that has five sides. The picture shows a regular pentagon.

Hexagon

A hexagon is a polygon that has six sides. The picture shows a regular hexagon.

Octagon

An octagon is a polygon that has eight sides. The picture shows a regular octagon.

Circle

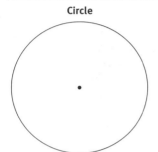

A circle is a figure on a two-dimensional surface in which all points are equidistant from a fixed point, called the center.

Ellipse

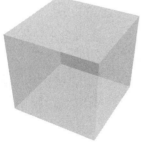

An ellipse is a figure on a two-dimensional surface in which the sum of the distances of the two foci to any point on the curve remains the same. (Note: *Foci* is plural; the singular form is *focus*.)

Cube

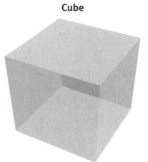

A cube is a three-dimensional figure with six identical square sides, called faces.

Cylinder

A cylinder is a three-dimensional figure with two bases that are congruent (identically sized) parallel circles.

Sphere 	A sphere is a three-dimensional figure in which all points are equidistant from a fixed point, called the center.
Pyramid 	A pyramid is a three-dimensional figure with a polygon base and triangular sides (or faces) that converge at a single point, called the vertex.
Tetrahedron 	A tetrahedron is a three-dimensional figure with four sides, each of which is a triangle.
Prism 	A prism is a three-dimensional figure with two parallel, congruent bases that are polygons. Prisms are named for the shape of their bases. For example, a triangular prism has triangles for bases, a rectangular prism has rectangles for bases, and a pentagonal prism has pentagons for bases.
Cone 	A cone is a three-dimensional figure with a circular base. The curved surface converges at a single point called the vertex. A cone is analogous to a pyramid with a circular base.

Index

Page references followed by *f* and *t* indicate figures and tables respectively.